Discontinuities in the Electromagnetic Field

IEEE Press
445 Hoes Lane
Piscataway, NJ 08854

Discontinuities in the Electromagnetic Field

M. Mithat Idemen

IEEE Antennas and Propagation Society, *Sponsor*

The IEEE Press Series on Electromagnetic Wave Theory
Andreas C. Cangellaris, *Series Editor*

IEEE Press

A John Wiley & Sons, Inc., Publication

Published simultaneously in Canada

Library of Congress Cataloging-in-Publication Data:

Idemen, M. Mithat.
Discontinuities in the electromagnetic field/M. Mithat Idemen.
p. cm.—(IEEE Press series on electromagnetic wave theory; 40)
ISBN 978-1-118-03415-6 (hardback)
1. Electromagnetic fields—Mathematics. 2. Maxwell equations. 3. Electromagnetic waves. I. Title.
QC665.E4I295 2011
530.14'1—dc22

2011002185

Printed in Singapore

oBook ISBN: 978-1-118-057902-6
ePDF ISBN: 978-1-118-05790-2
ePub ISBN: 978-1-118-05791-9

10 9 8 7 6 5 4 3 2 1

Contents

Preface

Without any hesitation we can claim that our civilization is largely based on the capacity and effectiveness of our knowledge pertaining to the natural phenomena collected under the name of *electromagnetism*. Our effective knowledge on these phenomena, with which we are familiar in fragments for nearly 2000 years, since the era of Thales, goes back only one and a quarter centuries ago. The years of 1873 and 1887 in the nineteenth century were two very important milestones in the intellectual evolution, as well as in the technological achievements of mankind. Indeed, the first one was the year in which the *electricity, magnetism*, and *optics*, which had been considered to be different natural phenomena until then, were unified under a common framework called the *electromagnetism*. To this end, Maxwell wrote a system of partial differential equations in England and claimed that those phenomena are all some particular aspects of a unique phenomenon which satisfies his equations. The equations introduced by Maxwell were not only what are reduced to the already known equations written 40 years ago by Faraday in England and 50 years ago by Ampère in France, but also claimed that the phenomenon in question consisted of a wave, which propagates with a finite velocity. What is very interesting is that physicists and engineers of those days were not familiar with such a kind of wave. To believe in the existence of the wave in question, one had to wait 14 years to witness the modest experiment carried out by Hertz in 1887 in Germany. That experiment, which revealed the so-called electromagnetic wave, showed also that energy transmission is possible through this wave. That was the indicator of a new direction for the civilization and thusly intensified the interest in the electromagnetic wave in both theoretical and experimental domains. Within the three years just after the Hertz experiment, the first radio receiver was realized by Branly in Germany in 1890, and soon after this, many crucial applications in communication, control, remote sensing, medicine, radio-astronomy, space communication, heating, and so on, began to flow. Among them we can mention, for example, the radio communication between two sides of the Atlantic Ocean (in 1901), radar (in 1940), laser (in 1957), tomography (in 1984), etc.

By considering new products of the actual technology, which permit us to use extremely short waves whose frequencies increase day by day, we can hope that electromagnetic waves will continue to be the basis of new, even unexpected, applications. These applications will require solutions of the Maxwell equations under new supplementary relations (boundary conditions, edge conditions, tip conditions, etc). Therefore, a thorough investigation of these supplementary relations is very important from both theoretical and practical points of view. The aim of this short monograph is to fulfill this job. I think (and hope) that readers who are fond of the *theory* will find several enjoyable points in the book. However, the meticulous colleagues will certainly find many points to criticize. As a scientist who imagines science as a continuously evolving struggle, I look forward with deep gratitude to all constructive comments and critiques.

What seems also important to me is that by claiming the existence of the electromagnetic wave before its experimental observation, the theory of Maxwell epitomizes an example for the "*theory before experiment.*" The electromagnetic *wave* is not the unique example presented by the Maxwell equations in this sense. Indeed, by claiming a nonuniversal time concept, they became also the main instigator for the Special Theory of Relativity established by Einstein nearly 30 years later than the Maxwell's Theory. The equations proposed by Maxwell conceal inside themselves too many secrets of the nature: reflected waves, refracted waves, creeping modes, whispering gallery modes, edge-excited waves, tip-excited waves, shadow boundaries, reflection boundaries, refraction boundaries, caustics, traveling waves, standing waves, Cherenkov waves, trapped waves, boundary conditions, edge conditions, tip conditions, space-time transformation under uniform motions, Doppler effect, aberration, and so on. They permit us to discover all these secretes as well as the interrelations among them if we interrogate with appropriate mathematical tools. I believe that the most effective tool for the investigation of the discontinuities in *natural phenomena* of any kind is the concept of *distribution*. This concept, which was introduced by Schwartz in 1950 (three quarters of a century later than the Maxwell equations) in France to the contemporary mathematics, extended the meaning of the Maxwell equations. If one assumes that the differential operations taking place in the Maxwell equations are all in the sense of distributions, then the discontinuities can be discussed pretty easily and rigorously. This monograph adopts this approach. When the surfaces which carry the discontinuities are in uniform motion, formulas of the Special Theory of Relativity help to reveal behaviors of the discontinuities by transforming the expressions pertinent to the surfaces at rest.

With their elegances and powers, the above-mentioned two theories (i.e., the Theory of Electromagnetism—including the special theory of relativity—and the Theory of Distributions) excite feelings of admiration. For me, they are not a heap of spiritless mathematical symbols but rather the *self-consistent* intellectual and *fine-artistic* productions of human mind. Since the Theory of Distributions as well as that of the Special Theory of Relativity are not thoroughly included into the undergraduate curriculum, in order to offer a self-contained book, the necessary (and sufficient) material which helps to clarify the essentials of these theories, are also be included into the book.

It is a great pleasure for me to confess that the publication of this monograph could not have been realized if sincere encouragements and supports by certain colleagues had not existed. Among them I mention especially Professor Tayfun Akgül (Istanbul Technical University, Turkey), Professor Ross Stone (IEEE Antennas and Propagation Society Publishing Board member), Professor Robert Mailloux (IEEE Antennas and Propagation Society Press Liason Committee Chair) and Professor Andreas C. Cangellaris (editor of the IEEE Press Series on Electromagnetic Wave Theory). I am indebted to all of them for their invaluable interests and supports.

I have also to thank Professor Alinur Büyükaksoy and Professor Ali Alkumru (both from the Gebze Institute of Technology, Turkey) for their several comments which helped me to improve the manuscript.

I express also my thanks to Taisuke Soda from the John Wiley and Sons, Inc. for his warm interest and meticulous work which produced a flawless publication, Lisa M. Van Horn of Wiley Production and Rajalakshmi Gnanakumar of Laserwords, for their superior support throughout the production process.

Lastly, I would like to extend my thanks and everlasting gratitude to TUBA (the Turkish Academy of Sciences) and Yeditepe University, Istanbul for partial support of my works.

Heybeliada, Istanbul M. MITHAT IDEMEN
March 2011

CHAPTER 1

Introduction

A man may imagine things that are false, but he can only understand things that are true.
Isaac Newton

Almost all mathematical problems connected with the electromagnetic phenomena require solutions of the Maxwell differential equations

$$\mathrm{curl}\mathbf{H} - \frac{\partial}{\partial t}\mathbf{D} = \mathbf{J}, \qquad \mathrm{curl}\mathbf{E} + \frac{\partial}{\partial t}\mathbf{B} \qquad (1.1\mathrm{a,b})$$

$$\mathrm{div}\mathbf{D} = \rho, \qquad \mathrm{div}\mathbf{B} = 0 \qquad (1.1\mathrm{c,d})$$

under certain supplementary restrictions stipulated on certain surfaces. In (1.1a–d) **E, D, H**, and **B** stand for the *electric field, electric displacement, magnetic field*, and *magnetic induction*, respectively, while ρ and **J** are the volume densities of the charges and currents. As to the parameter t that appears in (1.1a,b), it is, as usual, the time. The surfaces that bear the above-mentioned supplementary restrictions are either the interfaces between bodies of different constitutive parameters or surfaces that support surface charges and currents† or *material sheets* that model very thin layers. The supplementary restrictions in

*Michael White, *Isaac Newton: The Last Sorcerer*, Basic Books, New York, 1977, p. 5. See also *Sir Isaac Newton's Theological Manuscripts*, H. McLachlan (Ed.), Liverpool University Press, Liverpool, 1950, p. 17.

†Line and point charges are also carried by certain appropriately defined surfaces.

Discontinuities in the Electromagnetic Field, First Edition. By M. Mithat Idemen.

Published 2011 by John Wiley & Sons, Inc.

question are the relations that state the *physically admissible discontinuities* that may occur on these surfaces. They consist, in general, of the so-called *boundary conditions* that give the jump discontinuities on the surfaces in questions. If the discontinuity surfaces also involve sharp edges and/or sharp tips, then some components of the field become infinitely large at some points. In this case, in addition to the boundary conditions, one also has to know the physically admissible asymptotic behaviors of the field near those points because, as was shown more than 60 years ago by Bouwkamp [1], one can construct many solutions to the Maxwell equations under the given boundary conditions. Of course some of these solutions are not acceptable from physics point of view. Depending on the nature of the singular point, the relations that state the asymptotic behaviors in question are called the *edge conditions* or the *tip conditions*.

It is worthwhile to remark here that any relation written on a surface cannot be treated as a boundary condition for the electromagnetic field. In order to be so, it must also be compatible with the Maxwell equations. The spectrum of the electromagnetic waves used in the telecommunication is enlarging every day more and more toward very short waves. Hence many types of roughness, which had been assumed to be negligible in earlier investigations in order to reduce mathematical difficulties, became today unavoidable. In studies to be made in forthcoming days, one will have to consider the effects of these types of roughness on the propagation of waves. Therefore rigorous and detailed investigations of the boundary, edge, and tip conditions for various geometrical and physical structures are of crucial importance from both pure scientific and technological applications points of view. The aim of the present monograph is to study the discontinuities (i.e., singularities) in questions in their most general framework and discuss the validity of some particular relations that are in use in current literature. Our fundamental basis will be the so-called *distributions* (or *generalized functions*). In order to clarify the crucial role of this concept in the present study, it will be useful to reexamine the derivation of the *classical* boundary conditions in *almost* all textbooks.

On those days when the Electromagnetic Theory had been established, one had a huge heap of scientific knowledge in theoretical physics, especially in fluid mechanics. With this potential, the scientists of that time (i.e., mathematicians, physicists, and engineers) had formulated and solved many mathematical problems that could have interesting and important interpretations in terms of the electromagnetic notions. First problems were those that needed to find the explicit expressions of the fields created by various sources distributed in the vacuum. They were very easy. Soon later, one had considered the problems connected with bodies having simple geometrical shapes such as infinite planes, infinitely long cylinders, whole spheres, whole ellipsoids, and so on. They were also rather easy and tractable with known techniques provided that the boundary conditions to be satisfied on the surfaces of the bodies were known beforehand. It was at this stage that the struggles to reveal the discontinuities of the electromagnetic field were started. The methods that seemed most propitious were what are based on the applications of the Gauss–Ostrogradski and Stokes theorems (see Smirnow [2, pp. 177, 197]). Although these applications are repeated in almost all textbooks, we want to recapitulate them here in order to clarify the philosophy of the method adopted in the present study. To this end, consider, for example, (1.1c) and integrate it inside a volume ϑ bounded by planar boundaries of very small areas (see Fig. 1.1). Assume that $\mathbf{D}$ as well as its partial derivatives of the first orders are all *continuous inside* ϑ *except on a regular surface* ΔS. To avoid useless complexities, assume also that the field depends only on two space coordinates. Then a *careless* application (without regarding the validity conditions) of the Gauss–Ostrogradski theorem to

$$\int_{\vartheta} \operatorname{div}\mathbf{D}\, d\vartheta = \int_{\vartheta} \rho\, d\vartheta = Q \tag{1.2a}$$

yields

$$\int_{\Delta S_1+\Delta S_2+\Delta S_3+\Delta S_4} \mathbf{D}\cdot\mathbf{n}\, dS \qquad = Q. \tag{1.2b}$$

Here Q stands for the total charge existing inside ϑ. If Q consists of the surface charge distributed with density ρ_S on S, then by making

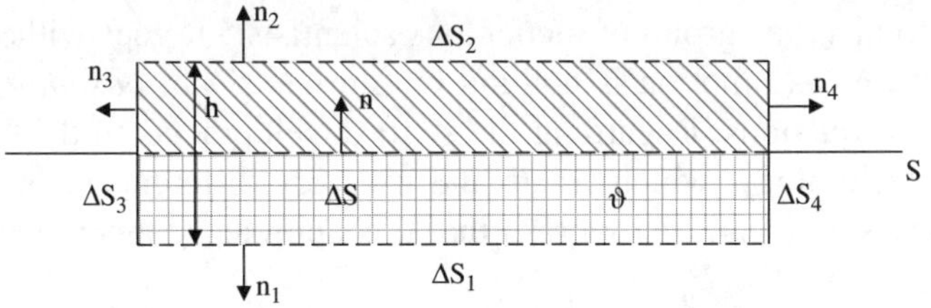

Figure 1.1. A neighborhood of a surface of discontinuity.

$h \to 0$ in (1.2b) one gets

$$\int_{\Delta S} [D_n^{(2)} - D_n^{(1)}]\, dS = \int_{\Delta S} \rho_S \, dS \tag{1.2c}$$

or

$$D_n^{(2)} - D_n^{(1)} = \rho_S. \tag{1.2d}$$

Here $D_n^{(1)}$ and $D_n^{(2)}$ show the values of the normal component of $\mathbf{D}$ which are observed when one approaches S from lower and upper sides, respectively.

Quite similarly, from (1.1a,b,d) one gets

$$B_n^{(2)} - B_n^{(1)} = 0, \tag{1.2e}$$

$$\mathbf{n} \times \mathbf{E}^{(2)} - \mathbf{n} \times \mathbf{E}^{(1)} = 0, \tag{1.2f}$$

$$\mathbf{n} \times \mathbf{H}^{(2)} - \mathbf{n} \times \mathbf{H}^{(1)} = \mathbf{J}_S. \tag{1.2g}$$

The field $\mathbf{J}_S$ appearing in (1.2g) is the density of the surface current that flows on S.

The relations in (1.2.d–g) are the *classical boundary conditions* that are satisfied on the interface between two regions filled with different materials. They are in use since the first days of the Theory to find solutions of the electromagnetic problems, which are in complete agreements with experiments. But the mathematical analysis made to reveal them is obviously *not legitimate*. Indeed, in order for the Gauss–Ostrogradski and Stokes theorems to be applicable, the fields **E, D, H**, and **B** have to be continuous inside ϑ. But, as the results themselves show, this is not the case. However, as we have already stated, the results are *correct from physics point of view*. A few writers who are familiar with this contrast warn readers by indicating that the results to be obtained by this kind of faulty applications are *assumed* to be acceptable for physics. D. S. Jones [3, p. 46] and S. A. Schelkunoff [4] epitomize this group of meticulous scientists.* It goes without saying that this assumption is nothing but an *additional postulate* to the Maxwell equations. Regarding the swift developments in the contemporary technology, which create and use very different materials and geometries, one can easily grasp that this approach cannot enable one

*Schelkunoff characterizes this kind of an approach to be "a proof which is not a proof but a swindle" (see Schelkunoff [4, Section 5]).

to overcome the difficulties completely. Hence one has to contrive a general and robust method.

One could also think that the difficulty in deriving (1.2d) resulted from the derivatives existing in (1.1c); if one started from the integral equation (Gauss' law)

$$\oint_S \mathbf{D} \cdot \mathbf{n}\, dS = Q, \tag{1.3a}$$

which is equivalent to (1.1c), one would not have the same difficulty. Indeed, in this case from (1.3a) one writes directly

$$\int_{\Delta S_1+\Delta S_2+\Delta S_3+\Delta S_4} \mathbf{D} \cdot \mathbf{n}\, dS = \int_{\Delta S} \rho_S\, dS, \tag{1.3b}$$

which, after applications of the Gauss–Ostrogradski theorem to the partial regions lying above and below the surface ΔS, yields (1.2d). But this approach too, which at first glance seems to be propitious to overcome the difficulty, has severe defects. For example, in the case when S consists of a double sheet that carries only dipoles, the right-hand side of (1.3b) becomes naught and claims $D_n^{(2)} - D_n^{(1)} = 0$, which is not correct.* Furthermore, in cases of material sheets that are represented, for example, by impedance or resistive or other more general type of boundary conditions one cannot guess the right-hand side of (1.3b).

To avoid the difficulties mentioned above, we propose to add the following assumption to the Maxwell equations [5]:

Maxwell equations are valid in the whole of the four-dimensional space in the sense of distribution.

The results of this assumption (or postulate) will be seen after Chapter 2. Here we confine ourselves to make only the following remarks that are of importance.

i. This assumption not only legitimizes the use of the Dirac delta *functions* to represent surface (or line or point) charges and currents (i.e., ρ and $\mathbf{J}$) but also claims that the field components $\mathbf{E}$,

*Correct expression is $\lambda[D_n^{(2)} - D_{1n}] = -\text{div}\{(1/\lambda)\rho_1\mathbf{n}\}$. Here ρ_1 stands for the density of the dipole distribution while λ is a parameter depending on the curvature of S and $\mathbf{n}$ is the unit normal vector to S [see Section 3.2.1, formula (3.19c)].

D, H, and **B** are also distributions. In other words, the field components themselves can contain *singular* terms concentrated on certain surfaces.

ii. This assumption concerns not only the space coordinates and boundary conditions but also the time parameter and initial values. That means that the field components can involve also singular terms concentrated at certain isolated instants (such as flash of lightning).

iii. In 1873, when Maxwell had established his theory, as well as during the following 75 years, the concept of distribution did not exist in the scientific literature. It appeared after 1950 first in mathematics and then in physics and engineering sciences. Hence, the claim that the Maxwell equations are valid in the sense of distribution is in fact a new postulate added to the Maxwell Theory.

The arrangement of the present monograph is as follows: In Chapter 2 the concepts of distribution and derivatives in the sense of distribution are explained. Grad, curl, and div operators as well as distributions concentrated on a surface are discussed in some detail. In Chapter 3 the Maxwell equations are reconsidered and discussed in the framework of these new concepts. The so-called *universal boundary conditions* are derived as a natural result. Chapter 4 is devoted to an extensive analysis of the boundary conditions on a *material sheet* at rest. The so-called impedance and resistive-type particular conditions are discussed and their validity conditions are derived. The case of moving boundaries is considered in Chapter 5. In this chapter the connection with the Special Theory of Relativity is also established whenever the motion is uniform. Chapter 6 is devoted to the edge conditions on a wedge bounded by planar walls. In this chapter, one shows also that the origin of the logarithmic-type singularities is the confluence of two algebraic singularities. In Chapter 7, one considers the tip singularities that occur at the apex of a rotationally curved material cone. For this kind of geometry, also the logarithmic singularities are derived as a result of confluence of two algebraic singularities. In Chapter 8, one considers the temporal discontinuities localized at certain times.

CHAPTER 2

Distributions and Derivatives in the Sense of Distribution

2.1 FUNCTIONS AND DISTRIBUTIONS

Even in the early days of the second half of the nineteenth century, one had observed that the mathematics, especially the concepts of functions and derivatives which had been established and maturated during the eighteenth and nineteenth centuries and permitted us to investigate various natural phenomena with deep insight, is not sufficient to grasp certain *singularities* in natural phenomena. To overcome the difficulties, one tried, from time to time, to introduce some concepts and to derive some formulas that were not based on solid mathematical basis. For example, Gustav Robert Kirchhoff,* a famous German physicist of the nineteenth century, had tried to define a force that acts on a very small area as follows [6]:

$$F = \left(\frac{\mu}{\sqrt{\pi}}\right) e^{-\mu^2 x^2} \qquad (\mu \text{ is a very large positive constant}). \qquad (2.1)$$

*G. R. Kirchhoff (Königsberg 1824–Berlin 1887).

Discontinuties in the Electromagnetic Field, First Edition. By M. Mithat Idemen.

Published 2011 by John Wiley & Sons, Inc.

For the same purpose Paul Dirac,* a very celebrated English physicist of the last century, had introduced his famous function† $\delta(x)$, which had the following properties [7]:

$$\delta(x) = 0 \qquad \text{when } x \neq 0 \tag{2.2a}$$

$$\delta(x) = \infty \qquad \text{when } x = 0 \tag{2.2b}$$

such that

$$\int_{-\infty}^{\infty} \delta(x)\, dx = 1. \tag{2.2c}$$

By using this exotic *δ-function*, Dirac had obtained some results that were adopted by physicists with enthusiasm. But the mathematicians of those days were watching the progress with certain reserve because they had known that the requirements stipulated by (2.2a)–(2.2c) were not compatible with the classical definitions of function and integral (in the Lebesgue sense).

Although it did not exist as a function in the proper sense of function, in the first half of the last century $\delta(x)$ was extensively used with its properties given in (2.2a–c) by physicists and engineers to produce many interesting results that could all be interpreted in an acceptable manner. This achievement encouraged mathematicians of that time to establish a rigorous basis for the exotic $\delta(x)$, which, from one perspective will ensure its adoption by mathematicians without any reserve and, from the other perspective, will permit one to interpret correctly the results obtained through it. This goal was achieved by the famous French mathematician Laurent Schwartz‡ in 1950 [8]. The result was the introduction of some new entities and concepts into the mathematics of the twentieth century. These new entities are the so-called *generalized functions*, which involve also all locally integrable functions. Considering previous uses of $\delta(x)$ to represent charge distributions, localized on point sets or lines or surfaces, these new entities were also referred to as the *distribution functions* or, more simply, *distributions*. In this monograph we will

*P. A. M. Dirac (Bristol 1902–Tallahassee 1984).

†The symbol δ had been used first by Kirchhoff and then later by Dirac (see Jones [3, p. 35]).

‡L. Schwartz (Paris 1915–2002). For this brilliant achievement he was honored by the Fields Medal in 1950. This award, which has been established in 1936 and devoted only to mathematics, is a counterpart of the Nobel Prize, which does not involve mathematics.

follow the founder of the theory and use this name—that is, *distribution*. In what follows we will recapitulate some basic notions and properties of distributions, which will constitute the main basis of our investigation.

2.2 TEST FUNCTIONS. THE SPACE C_0^∞

Let D be the set of real-valued functions that depend on the real variable $x \in (-\infty, \infty)$ such that beyond certain *finite* intervals they are naught and have continuous derivatives of all orders for all $x \in (-\infty, \infty)$. As we will see later on, these functions will play a crucial role in defining the distributions as well as their equalities, sums, multiplications, sequences, series, limits, derivatives, and so on. Hence the functions belonging to D are called the *test functions*. In what follows we want to clarify first the set D and its ability in the goal mentioned here.

To begin with, let us show that the set D is not empty. Indeed, the classical example given by Schwartz is as follows:

$$\varphi(x,a) = \begin{cases} \exp(-a^2/(a^2 - x^2)), & x \in [-a,a] \\ 0, & |x| \geq a. \end{cases} \tag{2.3}$$

The finite *closed* interval $[-a,a]$, outside of which $\varphi(x,a) \equiv 0$, is referred to as the *support** of $\varphi(x,a)$ and denoted by supp $\{\varphi(x,a)\}$ (see Fig. 2.1). A function having this property is called a function of *bounded support*. The set of test functions with support inside the interval (α, β) is also denoted by $C_0^\infty(\alpha,\beta)$, while $C_0(\alpha,\beta)$ shows the set of continuous functions with support inside (α,β). Note that $D = C_0^\infty(-\infty,\infty)$. We will denote D by C_0^∞ for short.

Starting from $\varphi(x,a)$ given in (2.3), we can also define many other test functions. Consider, for example, an arbitrary locally integrable† function $f(x) \in L_0^1(\alpha,\beta)$, which is identically naught beyond a certain

*Generally, the support of a function $f(x,y,\ldots)$ is the closure of the set of points $(x,y,\ldots)$ for which $f(x,y,\ldots) \neq 0$.

†If the integral of $|f(x)|$ on every finite interval exists, then $f(x)$ is said to be locally integrable.

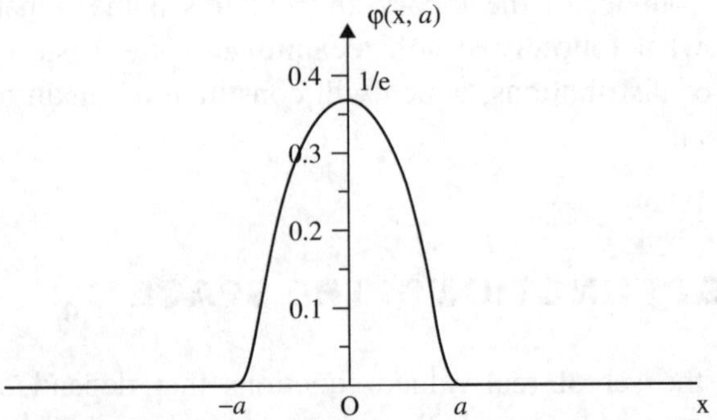

Figure 2.1. The test function $\varphi(x, a)$ and its support.

finite interval (α, β) and write

$$\varphi(x) = f(x) * \varphi(x, a) \tag{2.4a}$$

$$= \int_{\alpha}^{\beta} f(\xi)\varphi(x-\xi, a)\, d\xi \tag{2.4b}$$

$$= \int_{-a}^{a} f(x-\xi)\varphi(\xi, a)\, d\xi. \tag{2.4c}$$

In (2.4a), $*$ stands for the *convolution*. By considering the support of $f(x)$, we can easily convince ourselves that $f(x-\xi) \equiv 0$ when $x > (\beta + a)$ or $x < (\alpha - a)$ if $\xi \in (-a, a)$. Indeed, from $f(x) \equiv 0$ when $x > \beta$ one gets $f(x-\xi) \equiv 0$ when $x - \xi > \beta$ or $x > \beta + \xi$. Hence $f(x-\xi) \equiv 0$ for all $x > \beta + a$. Similarly, $f(x-\xi) \equiv 0$ for all $x < \alpha - a$. Thus from (2.4c) one concludes

$$\text{supp}\{\varphi(x)\} \subseteq (\alpha - a, \beta + a),$$

which claims that $\varphi(x)$ defined by (2.4a) is a function of bounded support (see Fig. 2.2). On the other hand, the expression in (2.4b) can be differentiated as many times as desired because the range of integration is finite and the integrand has continuous derivatives of all orders. This shows that the function $\varphi(x)$ is exactly a test function. By replacing $f(x)$ in (2.4a) by different functions, one gets different test functions. Two simple examples are shown in Fig. 2.2.

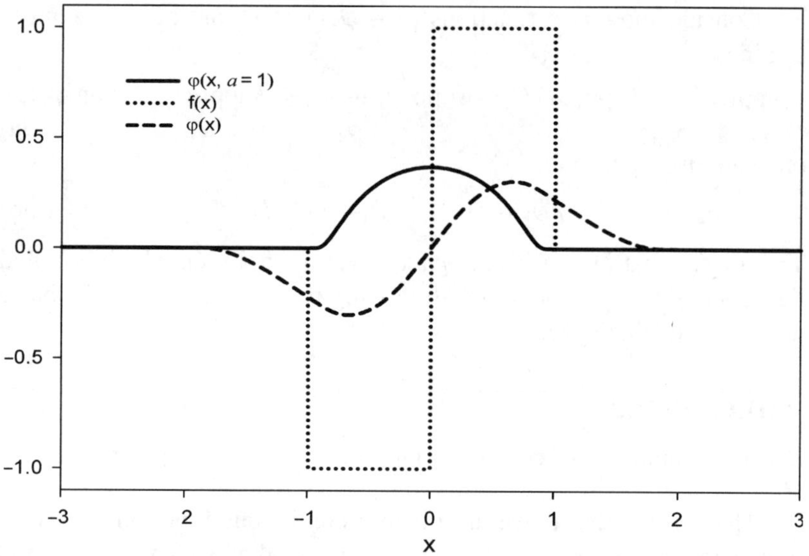

(a)

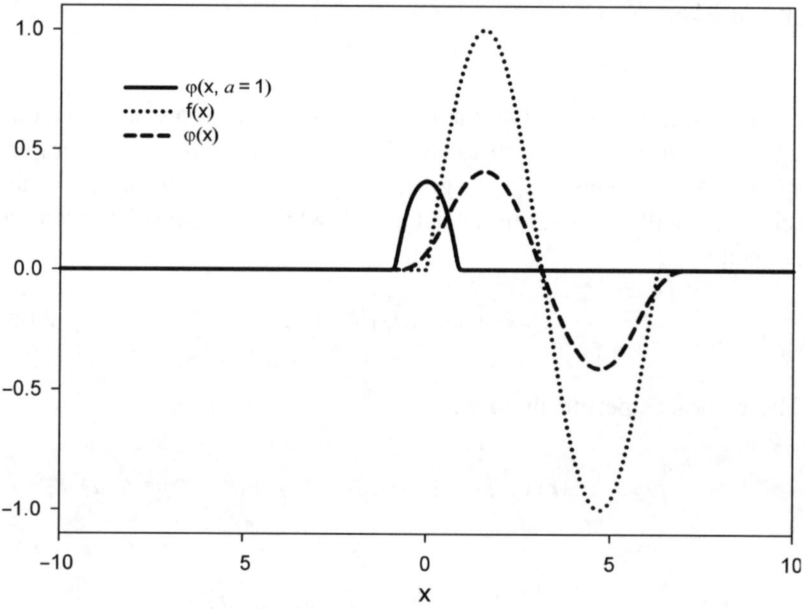

(b)

Figure 2.2. Test functions obtained through (2.4a). **(a)** An example for discontinuous $f(x)$. $f(x) \equiv 0$ when $|x| > 1$, $f(x) = -1$ when $x \in (-1, 0)$, and $f(x) = 1$ when $x \in (0, 1)$. **(b)** An example for continuous $f(x)$. $f(x) \equiv 0$ when $x \notin (0, 2\pi)$ while $f(x) = \sin x$ when $x \in (0, 2\pi)$.

Consider now two functions $\varphi_1 \in D, \varphi_2 \in D$ and a *real** number c. Since

$$\operatorname{supp}\{c\varphi_1\} = \operatorname{supp}\{\varphi_1\}, \qquad \operatorname{supp}\{\varphi_1 + \varphi_2\} \subseteq \operatorname{supp}\{\varphi_1\} \cup \operatorname{supp}\{\varphi_2\}, \tag{2.5a}$$

one can also claim that

$$c\varphi_1 \in D, \qquad \varphi_1 + \varphi_2 \in D. \tag{2.5b}$$

That means that D is a *linear space* (over the field R), which is called the *space of test functions*. The following two theorems state the basic properties of this space.

THEOREM 2.1

D is dense in the set of continuous functions of bounded support.

That means that if we are given a continuous function $f(x)$ with finite support, say (α, β), and a number $\varepsilon > 0$, then we can always find a function $\phi(\mathrm{x}) \in D$ such that the following inequality is satisfied for all $x \in (-\infty, \infty)$:

$$|f(x) - \phi(x)| < \varepsilon. \tag{2.6a}$$

In order to prove the theorem, we have to show at least one test function having this property. To this end consider, for example, $\phi(x) = A\varphi(x)$, where $\varphi(x)$ is given by (2.4a) with appropriately chosen a (a will be determined later on) while the constant $A > 0$ is defined by

$$A \int_{-a}^{a} \varphi(\xi, a)\, d\xi = 1. \tag{2.6b}$$

These choices permit us to write

$$f(x) - \phi(x) = Af(x) \int_{-a}^{a} \varphi(\xi, a)\, d\xi - A \int_{-a}^{a} f(x - \xi)\varphi(\xi, a)\, d\xi$$

$$= A \int_{-a}^{a} [f(x) - f(x - \xi)]\varphi(\xi, a)\, d\xi$$

*For more general studies, c may have also complex values. Here we will not need this kind of a generalization.

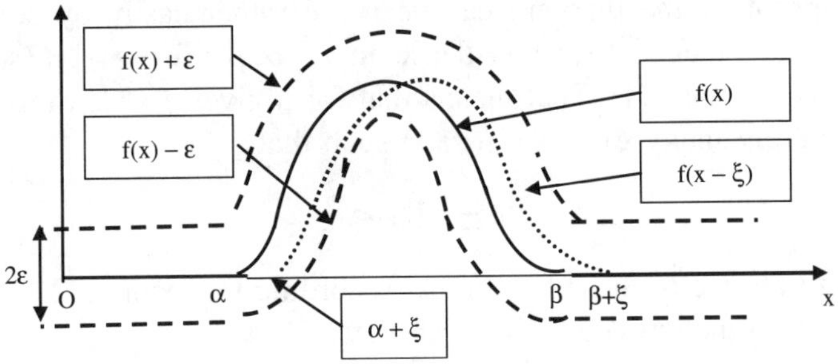

Figure 2.3. ε-neighborhood of $f(x)$.

or

$$|f(x) - \phi(x)| \leq A \int_{-a}^{a} |f(x) - f(x-\xi)| \varphi(\xi, a)\, d\xi. \tag{2.6c}$$

Now we choose $a = a(\varepsilon)$ such that $|f(x) - f(x-\xi)| < \varepsilon$ whenever $|x - (x - \xi)| = |\xi| < a$. We can do that because $f(x)$ is (uniformly!) continuous in the finite interval (α, β). To this end we translate first the graph of $f(x)$ upwards and downwards by ε to determine the ε-neighborhood of $f(x)$ (see Fig. 2.3). Then we slowly translate the same graph toward left and right by ξ to obtain the graph of $f(x - \xi)$. The maximum value of $|\xi|$, say $a_{\max}$, which yields a graph lying completely inside the above-mentioned ε-neighborhood of $f(x)$, determines the maximum value of a. We can choose any value smaller than $a_{\max}$ as a appearing in $\varphi(\xi, a)$. After having determined a in this manner, we determine also A through (2.6b). Thus (2.6c) becomes identical to (2.6a).

THEOREM 2.2

Let $L_0^p(\alpha, \beta)$ denote the set of functions $f(x)$ of support inside $(\alpha,\ \beta)$ such that

$$||f||_p = \left\{ \int_{\alpha}^{\beta} |f|^p dx \right\}^{1/p} < \infty, \qquad p = 1, 2, \ldots. \tag{2.7}$$

Then D is dense in $L_0^p(\alpha, \beta)$ in the sense that for every $f \in L_0^p(\alpha, \beta)$ and $\varepsilon > 0$ one can find $\phi \in D$ such that

$$||f - \phi||_p < \varepsilon, \qquad p = 1, 2, \ldots. \tag{2.8}$$

A proof of the theorem can be made rather easily by considering the fact that $C_0(\alpha,\beta)$ is dense in $L_0^p(\alpha,\beta)$ for $p \geq 1$ (see, for example, Adams [9]). That means that for a given $f \in L_0^p(\alpha,\beta)$ there are infinitely many $g(x) \in C_0(\alpha,\beta)$ such that

$$||f - g||_p < \frac{\varepsilon}{2}. \tag{2.9a}$$

Let $g(x)$ be one of these functions. According to Theorem 2.1 we can also find a function $\phi(x) \in D$ such that

$$|g(x) - \phi(x)| < \frac{\varepsilon}{2(\beta - \alpha)^{1/p}} \tag{2.9b}$$

is satisfied uniformly for all $x \in (-\infty,\infty)$. Thus we have

$$\begin{aligned} ||f - \phi||_p &= ||f - g + g - \phi||_p \\ &\leq ||f - g||_p + ||g - \phi||_p \\ &\leq \varepsilon/2 + \left[\int_\alpha^\beta |g - \phi|^p \, dx\right]^{1/p} \\ &\leq \varepsilon, \end{aligned}$$

which proves (2.8).

2.3 CONVERGENCE IN *D*

Consider now a sequence of test functions, say $\{\varphi_1, \varphi_2, \ldots\}$, and assume that the following requirements are met:

i. There exists a finite interval such that all φ_n as well as their derivatives of all orders—that is, $\varphi_n^{(m)}$ $(n = 1, 2, \ldots; m = 0, 1, \ldots)$—are naught outside this interval.

ii. When $n \to \infty$, all φ_n and $\varphi_n^{(m)}$ tend uniformly to zero.

Then one says that the sequence $\{\varphi_n\}$ *converges to zero in D*. Thus the above-mentioned conditions i and ii determine the topology of the space D. For example, all terms of the sequence

$$\varphi_n = \frac{1}{n}\varphi(x, a), \qquad n = 1, 2, \ldots, \tag{2.10a}$$

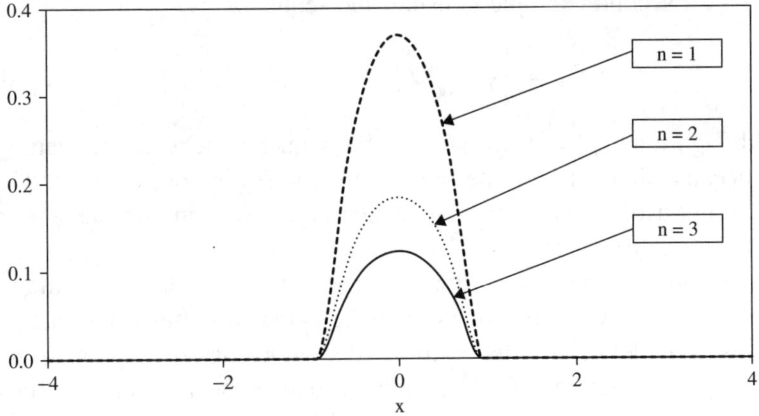

(**a**)

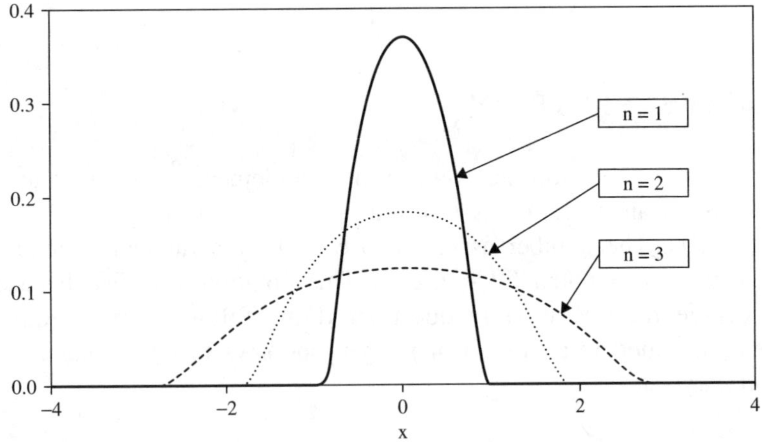

(**b**)

Figure 2.4. Sequences in D. **(a)** A sequence converging to zero. **(b)** A sequence not converging to zero.

where $\varphi(x, a)$ is defined through (2.3), are in D. All terms as well as their derivatives are obviously have the same support $(-a, a)$ (see Fig. 2.4a). Furthermore, one has also

$$\max |\varphi_n| = \frac{1}{ne} \to 0 \qquad \text{as } n \to \infty.$$

Hence this sequence converges to zero in D. To indicate this property we write $\varphi_n \to 0$.

As a second example consider the sequence

$$\tilde{\varphi}_n = \frac{1}{n}\varphi\left(\frac{x}{n}, a\right), \qquad n = 1, 2, \ldots \tag{2.10b}$$

Although max $|\tilde{\varphi}| = 1/(ne) \to 0$, this sequence does not converge in D because there is no finite interval mentioned in condition i. Indeed, the support of $\tilde{\varphi}_n$ is $(-na, na)$, which enlarges to infinity as $n \to \infty$ (see Fig. 2.4b).

Before leaving this section devoted to the presentation of the test functions, we want to remark that the set of test functions that can be used to define the distributions does not consist only of D. For some generalizations (for example, to define the Fourier transforms of distributions), one can also use a different set of test functions in which D is dense (see Gel'fand and Shilov [10]).

2.4 DISTRIBUTION

Consider an operation that assigns to each element of D a *real* number.* Such an operation is referred to as a *functional* defined in D (or on D). The number assigned to $\varphi \in D$ by a functional denoted by f will be denoted by $< f, \varphi >$. This notation will also be used to denote the operation in question. If the following two requirements are met by a functional f, then one says that f is *linear and continuous*:

i. For all $\varphi_1 \in D$ and $\varphi_2 \in D$, one has

$$< f,\ a\varphi_1 + b\varphi_2 > = a < f, \varphi_1 > + b < f, \varphi_2 > \quad \text{(linearity)}, \tag{2.11a}$$

where a and b are arbitrary real numbers.

ii. If φ_n tends to zero in D, then one has

$$< f, \varphi_n > \to 0, \qquad \text{as } n \to \infty \quad \text{(continuity)}. \tag{2.11b}$$

*If one considers complex-valued test functions and distributions, then the resulting functionals become also complex-valued. But in this monograph we omit this case that is a straightforward extension of the present case (see Section 2.5.3. Problem 1).

For example, if $f(x)$ is locally integrable in the sense of Riemann, then the functional defined through

$$<f,\varphi> = \int_R f(x)\varphi(x)\,dx \tag{2.12}$$

is linear and continuous.

Every linear and continuous functional defined on D is called a *generalized function* or a *distribution*. For example, the function $f(x)$ and the integration operation appearing in (2.12) define a distribution. It is obvious that every distribution cannot be defined through an integral.* The aforementioned δ-distribution of Dirac, which is defined by

$$<\delta,\varphi> = \varphi(0), \tag{2.13a}$$

epitomizes this case because $\delta(\mathrm{x})$ is not a function. However, in scientific literature one sees, rather frequently, notations like

$$<\delta,\varphi> = \int_R \delta(x)\varphi(x)\,dx. \tag{2.13b}$$

Although it is meaningless, this notation may sometimes be useful by playing role of guide when one tries to adapt some operations pertinent to functions to distributions. Among such operations we can mention, for example, translation, change of variables, derivation, and so on. However, one has to be careful in applying theorems and rules, established for functions, to integrals involving distributions. The set of distributions defined on D is denoted by D'. D' is called the *dual* of D.

The test functions belonging to D permit us to distinguish the elements of D'. Indeed, when we are given two distributions $f \in D'$ and $g \in D'$ such that

$$<f,\varphi> = <g,\varphi> \tag{2.14a}$$

for all $\varphi \in D$, then we say that these distributions are equal to each other. In such a case we write

$$f = g. \tag{2.14b}$$

*Note that by a theorem due to Riesz, every bounded linear functional in a Hilbert space can be defined through the inner product involving an *element of the space* determined by the functional in question (see, for example, Yoshida [11, p. 90]).

Owing to this separation property, the set D is called a *complete* space. To clarify the meaning of completeness mentioned here, let us consider two functions defined on the interval $(0, 2\pi)$, say $f(x)$ and $g(x)$, along with the sequence of trigonometric functions $\phi_n(x) = \cos nx\,(n = 0, 1, \ldots)$. It is very well known from the theorems pertinent to the Fourier series expansions [12] that from the identity

$$(f, \phi_n) = (g, \phi_n), \qquad n = 0, 1, \ldots,$$

where

$$(f, \phi_n) = \int_0^{2\pi} f(x)\phi_n(x)\,dx, \qquad (g, \phi_n) = \int_0^{2\pi} g(x)\phi_n(x)\,dx,$$

one *cannot* conclude the identity $f(x) \equiv g(x)$. But, if we replace the above-mentioned set of cosine functions—that is, $\{\phi_n(x)\}$—by $\{\psi_n(x)\} = \{1, \cos x, \sin x, \cos 2x, \sin 2x, \ldots\}$, then from the identity

$$(f, \psi_n) = (g, \psi_n), \qquad n = 0, 1, \ldots$$

one concludes $f(x) \equiv g(x)$ almost everywhere [12]. That means that in connection with the integrable functions defined on $(0, 2\pi)$ the sequence$\{\psi_n(x)\}$ is complete while the sequence $\{\phi_n(x)\}$ is not so.

It is worthwhile to remark that the claim of the completeness of D is based on the fact that when the distributions consist of locally integrable functions, as it was in (2.12), all the results to be obtained through D are compatible with already-proved results. Therefore, in order to prove the completeness of D, we have to show that the identity (2.14a) yields (2.14b) almost everywhere if $f(x)$ and $g(x)$ consist of *locally integrable* functions. To this end, consider a fixed interval (α, β) and define the function $F(x)$ as follows:

$$F(x) = \int_\alpha^x \{f(\xi) - g(\xi)\}d\xi, \qquad x \in (\alpha, \beta). \tag{2.15}$$

Since $f(x)$ and $g(x)$ are both integrable in (α, β), $F(x)$ is a continuous function of x. Therefore (2.14a) is equivalent to

$$\int_\alpha^\beta F'(x)\varphi(x)\,dx = 0, \qquad \forall\varphi \in C_0^\infty(\alpha, \beta). \tag{2.16a}$$

Hence, by the method of integration by parts, one writes also

$$0 = \int_{\alpha}^{\beta} F'(x)\varphi(x)\,dx = F(\beta)\varphi(\beta) - F(\alpha)\varphi(\alpha) - \int_{\alpha}^{\beta} F(x)\varphi'(x)\,dx$$

or

$$\int_{\alpha}^{\beta} F(x)\varphi'(x)\,dx = 0, \qquad \forall\varphi' \in C_0^{\infty}(\alpha,\beta). \tag{2.16b}$$

Here we have used the relations $F(\alpha) = 0, F(\beta) < \infty$, and $\varphi(\alpha) = \varphi(\beta) = 0$. From (2.16b) one gets $F(x) \equiv 0$ because $\varphi'(x)$ can be chosen very close to any continuous function. Indeed, if at a certain point $x_0 \in (\alpha,\beta)$ the continuous function $F(x)$ differs from zero, then without loss of generality we can assume that in a certain neighborhood $(x_0 - \delta, x_0 + \delta)$ of x_0 one has $F(x) > 0$. Then we can choose $\varphi'(x)$ in the ε neighborhood of the function $\psi(x) \in C_0(\alpha,\beta)$ given as follows (see Fig. 2.5):

$$\psi(x) = \begin{cases} 0 & \text{if } \alpha \le x \le x_0 - \delta \quad \text{or} \quad x_0 + \delta \le x \le \beta \\ 1 - (x - x_0)^2/\delta^2 & \text{if } x_0 - \delta \le x \le x_0 + \delta. \end{cases}$$

It is obvious that

$$\int_{\alpha}^{\beta} F(x)\psi(x)\,dx = \int_{x_0-\delta}^{x_0+\delta} F(x)\psi(x)\,dx = A > 0.$$

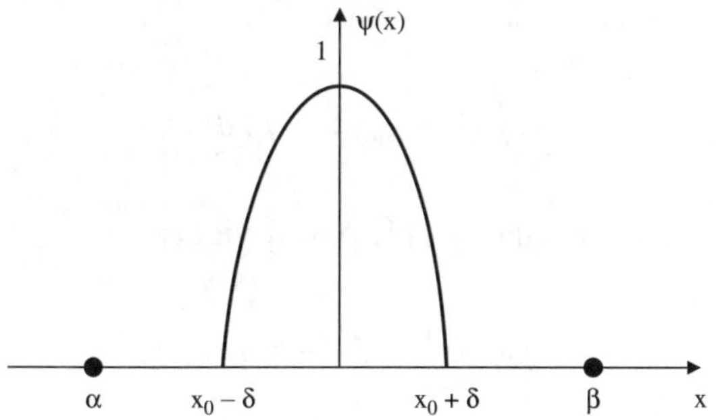

Figure 2.5. The function $\psi \in C_0(\alpha,\beta)$.

Now, from

$$\int_{\alpha}^{\beta} F(x)\varphi'(x)\,dx = \int_{\alpha}^{\beta} F(x)[\varphi'(x) - \psi(x)]\,dx + \int_{\alpha}^{\beta} F(x)\psi(x)\,dx$$

we write first

$$\left|\int_{\alpha}^{\beta} F(x)\varphi'(x)\,dx\right| \geq \int_{\alpha}^{\beta} F(x)\psi(x)\,dx - \left|\int_{\alpha}^{\beta} F(x)[\varphi'(x) - \psi(x)]dx\right|$$

$$\geq A - \varepsilon \int_{\alpha}^{\beta} |F(x)|dx$$

and then choose ε so small that the right hand-side of this relation becomes larger than $A/2 > 0$, which contradicts (2.16b). Therefore one has $F(x) \equiv 0$, which yields $F'(x) = f(x) - g(x) = 0$ almost everywhere in (α, β).

Remark. If f and g are in $L^2(\alpha, \beta)$, then (2.14b) can also be derived by the Schwarz inequality. Indeed, let $\phi \in L^2(\alpha, \beta)$ be any function while $\varphi \in C_0^{\infty}(\alpha, \beta)$ satisfies $||\phi - \varphi||_2 < \varepsilon$ for a given $\varepsilon > 0$ (see Theorem 2.2 above). Then, by considering also (2.14a), one writes

$$\int_{\alpha}^{\beta} (f - g)\phi\,dx = \int_{\alpha}^{\beta} (f - g)(\phi - \varphi + \varphi)\,dx$$

$$= \int_{\alpha}^{\beta} (f - g)(\phi - \varphi)\,dx + \int_{\alpha}^{\beta} (f - g)\,\varphi\,dx$$

$$= \int_{\alpha}^{\beta} (f - g)(\phi - \varphi)\,dx,$$

which by Schwarz inequality [11, p. 40] yields

$$\left|\int_{\alpha}^{\beta} (f - g)\phi\,dx\right| \leq ||f - g||_2 \cdot ||\phi - \varphi||_2$$

$$\leq ||f - g||_2 \varepsilon.$$

If in the latter we replace $\phi \in L^2(\alpha, \beta)$ by $(f - g) \in L^2(\alpha, \beta)$, then we get $||f - g||_2 \leq \varepsilon$ or, since ε is arbitrary, $f = g$ almost everywhere in (α, β).

2.5 SOME SIMPLE OPERATIONS IN *D′*

In our analysis we will very frequently apply some operations such as multiplication and derivation on distributions. Now we want to clarify how we extend these notions, established for ordinary functions, to the space of distributions.

2.5.1 Multiplication by a Real Number or a Function

Let $f \in D'$ be any distribution defined on R while $\psi \in C^\infty(-\infty, \infty)$ is any function (or constant) that has derivatives of all orders. Then the distribution (ψf) is defined by

$$< \psi f, \varphi > =< \mathrm{f}, \psi\varphi >, \qquad \varphi \in D. \tag{2.17}$$

It is important to observe that the requirements to be met by ψ are such that $(\psi\varphi) \in D$ whenever $\varphi \in D$. For example, $f = \delta$ yields

$$\begin{aligned} < \psi\delta, \varphi > &= < \delta, \psi\varphi > = \psi(0)\varphi(0) \\ &= \psi(0) < \delta, \varphi > =< \psi(0)\delta, \varphi > \end{aligned}$$

or

$$\psi(x)\delta(x) = \psi(0)\delta(x).$$

This shows that

$$\cos x\, \delta(x) = \delta(x), \qquad \sin x\, \delta(x) = 0$$

and more generally

$$x^n\delta(x) = 0 \quad \text{if } n = 1, 2, \ldots .$$

2.5.2 Translation and Rescaling

When $f(x)$ is a locally integrable function while a and b are two finite real numbers such that $a \neq 0$, we obviously have

$$< f(ax + b), \varphi(x) > = \int_R f(ax + b)\varphi(x)\, dx$$

$$= \int_R f(y)\varphi\left(\frac{y-b}{a}\right)\frac{dy}{|a|}$$

$$=< \frac{1}{|a|} f(x), \varphi\left(\frac{x-b}{a}\right) > . \tag{2.18a}$$

This relation hints how we have to define the translation and rescaling in the space of distributions. We will consider (2.18a) as the definition of these operations in D'. For example $\delta(x-b), \delta(ax)$ and $\delta(ax-b)$ will have the following meanings:

$$< \delta(x-b), \varphi(x) > =< \delta(x), \varphi(x+b) > = \varphi(b) \tag{2.18b}$$

$$< \delta(ax), \varphi(x) > =< \frac{1}{|a|}\delta(x), \varphi\left(\frac{x}{a}\right) > = \frac{1}{|a|}\varphi(0) \tag{2.18c}$$

and

$$< \delta(ax-b), \varphi(x) > =< \frac{1}{|a|}\delta(x), \varphi\left(\frac{x+b}{a}\right) > = \frac{1}{|a|}\varphi\left(\frac{b}{a}\right). \tag{2.18d}$$

2.5.3 Derivation of a Distribution

Let $f(x)$ and its first-order derivative $f'(x)$ be defined and continuous for all finite x. By taking into account the finiteness of the support of the test functions, we write by integration by parts

$$< f', \varphi > = \int_R f'(x)\varphi(x)\,dx = -\int_R f(x)\varphi'(x)\,dx = - < f, \varphi' > .$$

This result hints us how to extend the definition of derivative to the space of distributions. In accordance with it we state the following.

Definition. The derivative of a distribution f defined on R, say f', is the distribution defined as follows:

$$< f', \varphi > = - < f, \varphi' > . \tag{2.19}$$

Note that the right-hand side of (2.19) is already known because f is known. As a simple example, consider the unit step function of Heaviside, namely,

$$H(x) = \begin{cases} 1 & \text{when } x > 0, \\ 0 & \text{when } x < 0. \end{cases} \tag{2.20a}$$

From

$$< H', \varphi > = - < H, \varphi' > = - \int_0^\infty \varphi'(x)\, dx = \varphi(0) =< \delta, \varphi >$$

one concludes

$$H'(x) = \delta(x). \tag{2.20b}$$

It is worthwhile to remark here that $< H, \varphi' >$ was written as an integral because $H(x)$ is a locally integable function.

Now consider a function $f(x)$ that is continuous and locally integrable everywhere except $x = 0$, where it has a jump-type discontinuity as shown in Fig. 2.6. From the expression of $f(x)$ and (2.20b) one writes

$$f(x) = f_+(x)H(x) + f_-(x)[1 - H(x)] \tag{2.21a}$$

and

$$f'(x) = (b - a)\delta(x) + \{f'_+(x)H(x) + f'_-(x)H(-x)\}. \tag{2.21b}$$

This $f'(x)$ is referred to as the derivative of $f(x)$ in the sense of distribution. The term in the curly bracket denotes an ordinary function that is defined everywhere except $x = 0$ and gives the derivative in the classical sense. It is called the *regular part* of the distribution $f'(x)$ and is denoted by $\{f'\}$. As to the first term that appears with $\delta(x)$ and shows the jump of $f(x)$ at the point $x = 0$, it is said to be the *singular*

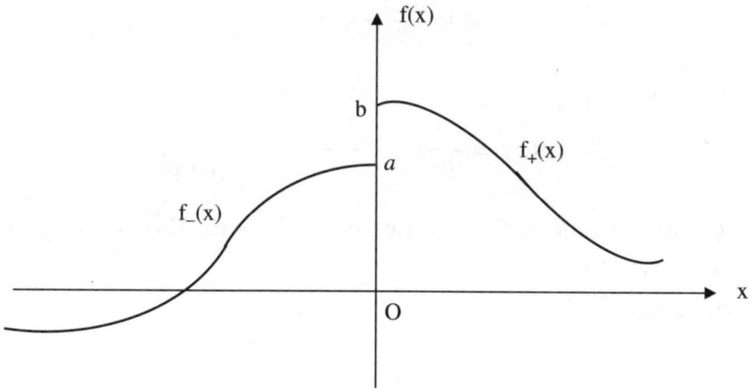

Figure 2.6. A discontinuous function with bounded jump.

part of the distribution $f'(x)$. In what follows, we will denote the finite discontinuity (jump) of $f(x)$ by $[[f]]$ and write (2.21b) as follows:

$$f' = \{f'\} + [[f]]\delta(x). \tag{2.21c}$$

For example, one has

$$\frac{d}{dx}[\sin xH(x)] = \sin x\,\delta(x) + \{\cos x\,H(x)\} = \cos x\,H(x)$$

$$\frac{d}{dx}[\cos xH(x)] = \cos x\delta(x) - \{\sin xH(x)\} = \delta(x) - \sin xH(x).$$

and

$$x\delta^{(n+1)}(x) = -(n+1)\delta^{(n)}(x),$$

$$-\frac{1}{6}x^3\delta'''(x) = \frac{1}{2}x^2\delta''(x) = -x\delta'(x) = \delta(x).$$

PROBLEMS

1. Let $z = x + iy$ and $f(z) = \log(z)$. Here $\log(z)$ denotes the principal branch of the logarithm function such that

$$\log(x + i0) = \log|x| + i\pi H(-x),$$
$$\log(x - i0) = \log|x| - i\pi H(-x).$$

(a) Show that

$$\lim_{y\to+0}\frac{d}{dx}\log(x+iy) = \lim_{y\to+0}\frac{1}{x+iy} = \frac{1}{x} - i\pi\delta(x).$$

(b) Use the second equality in (a) to show that

$$\lim_{y\to+0}\frac{y}{x^2+y^2} = \pi\delta(x).$$

(c) Show that

$$\delta(x) = \frac{1}{2\pi i}\left[\frac{1}{x-i0} - \frac{1}{x+i0}\right].$$

2. Let $f(x)$ be periodic with period equal to 2π such that in the interval $(0, 2\pi)$ one has $f(x) = (\pi - x)/2$.

(a) Show that

$$f'(x) = \{-1/2\} + \pi\sum_{-\infty}^{\infty}\delta(x - 2n\pi).$$

(b) Write the Fourier series expansion of $f'(x)$ and show that

$$f'(x) = \sum_{1}^{\infty} \cos nx.$$

(c) Compare the results in (a) and (b) to obtain

$$\sum_{-\infty}^{\infty} e^{inx} = 2\pi \sum_{-\infty}^{\infty} \delta(x - 2n\pi).$$

(d) Consider the functionals produced by both sides of the last equality and a test function $\varphi(x)$ to show that

$$\sum_{-\infty}^{\infty} \hat{\varphi}(n) = 2\pi \sum_{-\infty}^{\infty} \varphi(2n\pi) (\text{Poisson's formula}).$$

Here $\hat{\varphi}(\nu)$ stands for the Fourier transform of $\varphi(x)$, namely,

$$\hat{\varphi}(\nu) = \int_{-\infty}^{\infty} e^{i\nu x} \varphi(x)\, dx.$$

(e) From the latter conclude that

$$\sqrt{\beta} \left\{ \frac{1}{2} F_c(0) + F_c(\beta) + F_c(2\beta) + \cdots \right\}$$
$$= \sqrt{a} \left\{ \frac{1}{2} f(0) + f(\alpha) + f(2\alpha) + \cdots \right\},$$

where $\alpha > 0$ and $\beta > 0$ are given constants such that $\alpha\beta = 2\pi$ while $F_c(\omega)$ stands for the cosine transform of $f(x) \in C[0, \infty)$ defined as follows:

$$F_c(\omega) = \sqrt{\frac{2}{\pi}} \int_{0}^{\infty} f(\xi) \cos(\xi\omega)\, d\xi.$$

3. Let $y(x)$ satisfy the differential equation $y'' + 4y' + 4y = |x|$ in the sense of distribution under the two-point boundary conditions $y(-1) = 0$ and $y(1) = 0$.

(a) Show that one has also

$$y(+0) = y(-0), \qquad y'(+0) = y'(-0).$$

(b) Find an explicit expression of $y(x)$.

4. Let $G(x,\xi)$ satisfy $G_{xx} + k^2 G = \delta(x-\xi)$ under the boundary conditions $G(0,\xi) = 0$ and $G(\pi,\xi) = 0$, where ξ stands for a parameter in the interval $(0, \pi)$.

(a) Find an explicit expression of $G(x,\xi)$ and show that $G(x,\xi) = G(\xi,x)$. Discuss the case where k is equal to an integer.

(b) Show that the solution to the boundary-value problem

$$u'' + k^2 u = f(x), \quad x \in (0,\pi); \qquad u(0) = 0, \quad u(\pi) = 0$$

is

$$u(x) = \int_0^{\pi} G(x,\xi) f(\xi)\, d\xi.$$

Here $f(x) \in L^1(0,\pi)$ is an arbitrary function. [$G(x,\xi)$ is referred to as the Green's function associated with the boundary-value problem mentioned in (b)].

5. Show that one always has

$$\frac{\delta'(x-\alpha)}{x} - \frac{\delta'(x-\alpha)}{\alpha} = \frac{\delta(x-\alpha)}{\alpha^2}$$

$$\frac{\delta'(x-\alpha)}{x^2} - \frac{\delta'(x-\alpha)}{\alpha^2} = 2\frac{\delta(x-\alpha)}{\alpha^3}$$

$$\frac{\delta'(\theta-\psi)}{\sin\theta} - \frac{\delta'(\theta-\psi)}{\sin\psi} = \frac{\cos\psi}{\sin^2\psi}\delta(\theta-\psi)$$

and, more generally,

$$[\psi(x) - \psi(\alpha)]\delta'(x-\alpha) = -\psi'(\alpha)\delta(x-\alpha).$$

2.6 ORDER OF A DISTRIBUTION

By considering the fact that C_0^∞ consists of functions that are infinitely many times differentiable, in $C_0^\infty(\alpha,\beta)$ one can define a sequence of norms, say $||\cdot||_p$, by*

$$||\varphi||_p = \max_{\alpha \le x \le \beta} \{|\varphi|, |\varphi'|, \ldots, |\varphi^{(p)}|\}, \qquad p = 0, 1, \ldots. \tag{2.22a}$$

It is obvious that $||\varphi||_p$ meets all axioms put on the norm and that the sequence $\{||\cdot||_p\}$ is nondecreasing, namely,

$$||\varphi||_p \le ||\varphi||_{p'}, \qquad p < p'. \tag{2.22b}$$

*Note the difference between $||\cdot||_p$ defined in (2.7) and (2.22a).

Indeed, to check the triangle inequality we can write

$$\begin{aligned} ||\varphi_1 + \varphi_2||_p &= \max_{\alpha \le x \le \beta} \{|\varphi_1 + \varphi_2|, \ldots, |\varphi_1^{(p)} + \varphi_2^{(p)}|\} \\ &\le \max_{\alpha \le x \le \beta} \{|\varphi_1| + |\varphi_2|, \ldots, |\varphi_1^{(p)}| + |\varphi_2^{(p)}|\} \\ &\le \max_{\alpha \le x \le \beta} \{|\varphi_1|, \ldots, |\varphi_1^{(p)}|\} + \max_{\alpha \le x \le \beta} \{|\varphi_2|, \ldots, |\varphi_2^{(p)}|\} \\ &\le ||\varphi_1||_p + ||\varphi_2||_p. \end{aligned}$$

Now assume that for a distribution f one can find a *finite* constant $C > 0$ and an integer $p \ge 0$ such that

$$| < f, \varphi > | \le C||\varphi||_p, \qquad \forall \varphi \in C_0^\infty(\alpha, \beta). \tag{2.23a}$$

In this case, one says that f is a distribution of *finite order in the interval* (α, β). If in (2.23a) p is replaced by a larger integer p', then, by virtue of (2.22b) it remains still valid with the same C. But, on the contrary, when p is replaced by a smaller integer p', (2.23a) may not be valid with the same C because the second factor on the right-hand side of it becomes reduced. Nevertheless, for certain values of p', (2.23a) may still be written with different (*finite*) constants C' larger than C. The least value of p', say k, for which a relation of the form (2.23a) can be written with a finite constant C is referred to as the *order of the distribution f in the interval* (α, β) (cf. [10, p. 34]). Thus one writes

$$| < f, \varphi > | \le C||\varphi||_k, \qquad \forall \varphi \in C_0^\infty(\alpha, \beta). \tag{2.23b}$$

It is worthwhile to remark here that in different intervals f may be of different orders. For example, the Dirac distribution $\delta(x)$ is equal to naught in every interval not including the point $x = 0$ while it is of order zero in every interval including the point $x = 0$ (see the remark and examples to be given at the end of this section).

Distributions of finite orders have a very important property stated by the following theorem.

THEOREM 2.3 (L. Schwartz)

Let f be a distribution of order k in an interval (α, β). Then there exists a function $F(x) \in C(\alpha, \beta)$ such that $f = F^{(k+2)}(x)$ in the sense of distributions.*

*Note that $F(x)$ is not unique. See Remark 2 at the end of this subsection.

Proof (Cf. Gel'fand and Shilov [10, pp. 109–119]
We will prove this theorem by reducing f into a bounded functional defined in a *certain* Hilbert space and using the Riesz theorem pertinent to such functionals. To this end, use (2.22a) in (2.23b) and write

$$|< f, \varphi >| \leq C \max_{\alpha \leq x \leq \beta} \{|\varphi|, |\varphi'|, \ldots, |\varphi^{(k)}|\}, \qquad \forall \varphi \in C_0^{\infty}(\alpha, \beta). \tag{2.24}$$

Now consider the function $\varphi^{(k+1)}(x)$ and the mapping $\varphi \leftrightarrow \varphi^{(k+1)} \equiv \psi$. The latter is a bijection. Indeed, the uniqueness in the direction $\varphi \to \psi$ is obvious. To see the uniqueness in the inverse direction (i.e., for $\psi \to \varphi$), note that from $\psi \to \varphi_1$ and $\psi \to \varphi_2$ one writes first $\varphi_1^{(k+1)} - \varphi_2^{(k+1)} \equiv 0$, which yields $\varphi_1^{(k)} - \varphi_2^{(k)} \equiv \text{constant} = C$. Then, by considering the fact that $\text{supp}\{\varphi_{1,2}\} \subseteq [\alpha, \beta]$, one gets $C = \varphi_1^{(k)}(\alpha) - \varphi_2^{(k)}(\alpha) = 0$. By repeating this process $(k+1)$ times, we get finally $\varphi_1 \equiv \varphi_2$. Hence, $< f, \varphi >$ can also be thought as the value of a certain functional, say T, applied to ψ, namely,

$$< T, \psi > \equiv < f, \varphi > . \tag{2.25}$$

In other words, (2.25) completely defines the functional T because to find the value of $< T, \psi >$ we find first φ and then compute $< f, \varphi >$. On the other hand, we can show that $< T, \psi >$ is a *bounded linear functional* in the Hilbert space $L^2[\alpha, \beta]$ (we will show this later on; see (2.32a)). Therefore, by the Riesz Theorem it can be written as [11, p. 90]

$$< T, \psi > \equiv (F_1, \psi) = \int_a^{\beta} F_1(x)\psi(x)dx, \tag{2.26}$$

where $F_1(x) \in L^2[\alpha, \beta] \subset L^1[\alpha, \beta]$ stands for a suitable function to be determined by T (or f).

Let the integral of $F_1(x) \in L^1[\alpha, \beta]$ be $F(x)$, namely,

$$F(x) = \int_{\alpha}^{x} F_1(\lambda)\, d\lambda.$$

$F(x)$ is a continuous function in $[\alpha, \beta]$. Thus from (2.25) and (2.26) we write

$$<f,\varphi> =< T,\psi> = (F_1,\psi) = (F',\varphi^{(k+1)}) \\ =< (-1)^{k+1}F^{(k+2)},\varphi>, \quad (2.27a)$$

which yields

$$f = (-1)^{k+1}F^{(k+2)}(x) \qquad \text{(in the sense of distribution).} \quad (2.27b)$$

To complete the proof, we have to show finally that $< T, \psi >$ is a bounded linear functional in $L^2[\alpha, \beta]$, namely,

$$| < T,\psi > | \leq K||\psi||_{L^2[\alpha,\beta]}, \quad (2.28)$$

where K refers to a certain constant not depending on ψ. To this end, we start from

$$\varphi(x) = \int_\alpha^x \varphi'(\lambda)d\lambda$$

and write, by Schwarz inequality [11, p. 40],

$$|\varphi(x)| \leq \int_\alpha^x |\varphi'(\lambda)|d\lambda \\ \leq \int_\alpha^\beta |\varphi'(\lambda)|d\lambda \\ \leq \{\int_\alpha^\beta 1^2 d\lambda\}^{1/2}\{\int_\alpha^\beta |\varphi'(\lambda)|^2 d\lambda\}^{1/2}.$$

This shows that for $p \in \{0, 1, \ldots, k\}$ one has

$$|\varphi^{(p)}(x)| \leq (\beta-\alpha)^{1/2}||\varphi^{(p+1)}||_{L^2} \quad (2.29)$$

or

$$|\varphi^{(p)}(x)|^2 \leq (\beta-\alpha)||\varphi^{(p+1)}||_{L^2}^2$$

and

$$\int_\alpha^\beta |\varphi^{(p)}(x)|^2 dx \leq (\beta-\alpha)^2||\varphi^{(p+1)}||_{L^2}^2,$$

which gives

$$||\varphi^{(p)}(x)||_{L^2} \leq (\beta - \alpha)||\varphi^{(p+1)}||_{L^2}, \qquad p = 0, 1, \ldots, k. \tag{2.30}$$

If one uses (2.30) in (2.29) successively, then one arrives at

$$\begin{aligned} |\varphi^{(p)}(x)| &\leq (\beta - \alpha)^{1/2+1}||\varphi^{(p+2)}||_{L^2} \\ &\leq (\beta - \alpha)^{1/2+2}||\varphi^{(p+3)}||_{L^2} \\ &\vdots \\ &\leq (\beta - \alpha)^{1/2+(k-p)}||\varphi^{(k+1)}||_{L^2}. \end{aligned} \tag{2.31}$$

The use of (2.31) in (2.24) and (2.25) yields

$$\begin{aligned} &| < T, \psi > | \equiv | < f, \varphi > | \leq C \max_{\alpha \leq x \leq \beta} \{|\varphi|, |\varphi'|, \ldots, |\varphi^{(k)}|\} \\ &\leq C \max_{\alpha \leq x \leq \beta} \{(\beta - \alpha)^{1/2+k}||\varphi^{(k+1)}||_{L^2}, \ldots, (\beta - a)^{1/2}||\varphi^{(k+1)}||_{L^2}\} \\ &\leq K||\varphi^{(k+1)}||_{L^2} \\ &\leq K||\psi||_{L^2}, \end{aligned} \tag{2.32a}$$

where one put

$$K = \begin{cases} C(\beta - a)^{(k+1/2)} & \text{if } (\beta - a) > 1 \\ C(\beta - a)^{1/2} & \text{if } (\beta - \alpha) < 1. \end{cases} \tag{2.32b}$$

Equation (2.32a) asserts the boundedness of the functional T in $L^2[\alpha, \beta]$. ■

EXAMPLE

For every $\varphi \in C_0^\infty[-a, a]$ the Dirac distribution $\delta(x)$ gives $< \delta, \varphi > = \varphi(0)$. This shows that for every interval about the origin, one has

$$| < \delta, \varphi > | = |\varphi(0)| \leq \max_{-a \leq x \leq a} \{|\varphi(x)|\},$$

whence one concludes $k = 0$ and

$$\delta(x) = \frac{d^2}{dx^2} F(x), \qquad F(x) \in C[-a, a].$$

One can easily check that the following function can be thought as $F(x)$ mentioned here:

$$F(x) = \frac{1}{2}|x|.$$

Remarks

1. It is worthwhile to remark that for an interval $[\alpha, \beta]$ not involving the origin, one always has

$$< \delta, \varphi > = 0, \qquad \forall \varphi \in C_0^\infty[\alpha, \beta].$$

This shows that in $[\alpha, \beta]$ one can write $\delta \equiv 0 \in C[\alpha, \beta]$. In other words, $\delta(x)$ coincides with a continuous function (null function). This corresponds to $k = -2$ in (2.27b).

2. For $\delta(x)$ one can also write $\delta(x) = F''(x)$ with

$$F(x) = \begin{cases} 0, & x<0 \\ x, & x > 0. \end{cases}$$

This shows that the function $F(x)$ mentioned in the statement of Theorem 2.3 is not unique. This is quite plausible because the relation $f(x) = F^{(k+2)}(x)$ constitutes a differential equation for $F(x)$, which may have many solutions. ∎

EXAMPLE 1

Let $f = \delta' + \delta + H(x)$. Here $H(x)$ refers, as usual, to the unit step function of Heaviside. Then show that $k = 1$ and $f = F'''(x)$ with

$$F(x) = \begin{cases} x + \frac{1}{2}x^2 + \frac{1}{6}x^3 & \text{when } x \geq 0 \\ 0 & \text{when } x \leq 0. \end{cases}$$

2.7 THE SUPPORT OF A DISTRIBUTION

Definition of the support of a function such as $\phi \in C_0(\alpha, \beta)$ is a simple matter. But the job is rather hard for a distribution whose values at discrete points are not defined, in general. Now we want to extend the definition given in Section 2.2 to distributions.

Definition. Let $I \subset R$ be a given set while f is a given distribution on R. If $< f, \varphi > = 0$ for all $\varphi \in D$ with support outside I, then we say that the support of f is inside the set I. The intersection of all sets I having this property is referred to as the support of f.

For example,

$$\text{Supp}\{\delta(x)\} = \{0\}. \tag{2.33}$$

THEOREM 2.4

Let f be a distribution of finite order defined on R. If the support of f consists only of the point $x = 0$, then it is of the following form:

$$f = c_0\delta(x) + c_1\delta'(x) + \cdots + c_k\delta^{(k)}(x). \tag{2.34a}$$

Here k stands for the order of f while $c_j\,(j = 0, 1, \ldots, k)$ are certain (real or complex-valued) constants.

Proof

This is in fact a direct corollary of Theorem 2.3. Indeed, from (2.27a) one writes

$$< F^{(k+2)}, \varphi > = (-1)^{k+1} < f, \varphi > = 0$$

for all test functions with support in the region $x > 0$. This shows that $F^{(k+2)}(x) \equiv 0$ and, consequently, the continuous function $F(x)$ is a polynomial of degree $(k+1)$ in the region $x > 0$, namely,

$$F(x) = b_0 + b_1 x + \cdots + b_{k+1}x^{k+1}, \qquad x > 0. \tag{2.34b}$$

By a similar reasoning, one writes in the region $x < 0$

$$F(x) = a_0 + a_1 x + \cdots + a_{k+1}x^{k+1}, \qquad x < 0. \tag{2.34c}$$

From the continuity of $F(x)$ one gets $a_0 = b_0$. If one computes the derivatives of $F(x)$ in the sense of distribution by considering (2.21c), then from (2.34b) and (2.34c) one obtains

$$F'(x) = \begin{cases} b_1 + 2b_2 x + \cdots + (k+1)b_{k+1}x^k, & x > 0, \\ a_1 + 2a_2 x + \cdots + (k+1)a_{k+1}x^k, & x < 0, \end{cases}$$

$$F''(x) = (b_1 - a_1)\delta(x)$$

$$+ \begin{cases} 2b_2 + 3.2b_3 x + \cdots + k(k+1)b_{k+1}x^{k-1}, & x > 0 \\ 2a_2 + 3.2a_3 x + \cdots + k(k+1)a_{k+1}x^{k-1}, & x < 0 \end{cases}$$

and

$$F^{(k+2)}(x) = (b_1 - a_1)\delta^{(k)}(x) + 2(b_2 - a_2)\delta^{(k-1)}(x)$$

$$+ \cdots + (k+1)!(b_{k+1} - a_{k+1})\delta(x). \tag{2.34d}$$

The latter is nothing but (2.34a). ■

2.8 SOME GENERALIZATIONS

The subject of the present monograph involves functions of several variables and, hence, needs to generalize the concepts defined above to entities depending on several variables. Now we will examine some generalizations that we will need later on.

2.8.1 Distributions on Multidimensional Spaces

All the test functions and distributions we have considered so far were defined on a one-dimensional space, more precisely on R. All definitions made there as well as all results can be extended by straightforward generalizations to the case of distributions defined on multidimensional spaces. For example, the test function $\varphi(x,a)$ given in (2.3) is replaced in R^3 by

$$\varphi(x_1,x_2,x_3,a) = \begin{cases} \exp\{-a^2/(a^2-|x|^2)\}, & |x| \le a \\ 0, & |x| \ge a, \end{cases} \tag{2.35a}$$

with

$$|x| = \sqrt{(x_1)^2 + (x_2)^2 + (x_3)^2}. \tag{2.35b}$$

It is obvious that $\varphi(x_1,x_2,x_3,a)$ is naught outside the sphere of radius a and has continuous partial derivatives of all orders for all $x \in R^3$.

Similarly, the distributions defined by (2.12) and (2.13a) are extended to R^3 as follows:

$$< f(x,y,z), \varphi(x,y,z) > = \iiint_{R^3} f\varphi dxdydz \tag{2.36}$$

$$< \delta(x,y,z), \varphi(x,y,z) > = \varphi(0,0,0). \tag{2.37}$$

Multiplication of two distributions defined on the same space (i.e., depending on the same coordinates) is not defined. For example $\delta^2(x)$ has no meaning. But the product of distributions depending on different coordinates is treated by straightforward generalization, namely,

$$\begin{aligned} &< f_1(x)f_2(y)f_3(z), \varphi(x,y,z) > \\ &=< f_1(x), < f_2(y), < f_3(z), \varphi(x,y,z) \gg. \end{aligned} \tag{2.38}$$

Hence the distribution $\delta(x, y, z)$ defined by (2.37) can also be written as follows:

$$\delta(x, y, z) = \delta(x)\delta(y)\delta(z).$$

Let S be a regular* surface. The distribution $\delta(S)$ *concentrated* on S (i.e., with support S) is defined by

$$< \delta(S), \varphi(x, y, z) > = \iint\limits_{S} \varphi(x, y, z)\, dS. \tag{2.39a}$$

In accordance with this definition, the aforementioned distribution $\delta(x)$ defined on R can also be treated as a distribution concentrated on the plane $x = 0$ in R^3, namely,

$$< \delta(x), \varphi(x, y, z) > = \int\limits_{y=-\infty}^{\infty} \int\limits_{z=-\infty}^{\infty} \varphi(0, y, z)\, dydz. \tag{2.39b}$$

In multidimensional spaces, one-to-one coordinate transformations are applied as usual. When the transformation is not one-to-one, the space is appropriately divided into subspaces such that in each of them the transformation is one-to-one. In order to clarify the subject, we want to give three examples.

EXAMPLE 1

Consider the transformation from Cartesian to circular cylindrical coordinates in R^2, namely,

$$x = r\cos\theta, \qquad y = r\sin\theta, \tag{2.40a}$$

where $r \in [0, \infty)$ is the radial distance (from the origin) while $\theta \in [0, 2\pi)$ is the polar angle. To find the expression of the distribution

$$\delta(x - x_0, y - y_0) = \delta(x - x_0)\delta(y - y_0)$$

in terms of the polar coordinates r and θ, say $\tilde{\delta}(r, \theta)$, we consider the definitions as follows:

$$\begin{aligned}\varphi(r_0\cos\theta_0, r_0\sin\theta_0) &\equiv \varphi(x_0, y_0)\\ &=< \delta(x - x_0, y - y_0), \varphi(x, y) >\end{aligned}$$

*At every point of S, there is a unique tangent plane which varies continuously from point to point.

$$= \iint\limits_{R} \delta(x - x_0)\delta(y - y_0)\varphi(x,y)\, dxdy$$

$$= \int\limits_0^{\infty}\int\limits_0^{2\pi} \tilde{\delta}(r,\theta)\varphi(r\cos\theta, r\sin\theta) r d\theta dr.$$

Here r_0 and θ_0 denote the polar coordinates of the point (x_0, y_0). From the identity of the first and last terms, one gets

$$\tilde{\delta}(r,\theta) = \frac{1}{r}\delta(r - r_0)\delta(\theta - \theta_0). \tag{2.40b}$$

This shows that the transformation from the Cartesian coordinates (x, y) to the polar coordinates (r, θ) yields

$$(x,y) \to (r,\theta), \qquad \text{dxdy} \to r dr d\theta,$$

and

$$\delta(x - x_0, y - y_0) \to \frac{1}{r}\delta(r - r_0)\delta(\theta - \theta_0). \tag{2.40c}$$

EXAMPLE 2

Consider now the similar transformation in R^3 and write

$$x = r\sin\theta\cos\phi, \qquad y = r\sin\theta\sin\phi, \qquad z = r\cos\theta, \tag{2.41a}$$

where $r \in [0, \infty), \theta \in [0, \pi]$, and $\phi \in [0, 2\pi)$. By repeating the discussion made above, one writes first

$$\begin{aligned}
&\varphi(r_0\sin\theta_0\cos\phi_0, r_0\sin\theta_0\sin\phi_0, r_0\cos\theta_0)\\
&\quad \equiv \varphi(x_0, y_0, z_0)\\
&\quad =< \delta(x - x_0, y - y_0, z - z_0), \varphi(x,y,z) >\\
&\quad = \iiint \delta(x - x_0)\delta(y - y_0)\delta(z - z_0)\varphi(x,y,z) dxdydz\\
&\quad = \int\limits_0^{\infty}\int\limits_0^{2\pi}\int\limits_0^{\pi} \tilde{\delta}(r,\theta,\phi)\varphi(r\sin\theta\cos\phi, r\sin\theta\sin\phi, r\cos\theta) r^2 \sin\theta d\theta d\phi dr
\end{aligned}$$

and then obtain

$$\tilde{\delta}(r,\theta,\phi) = \frac{1}{r^2\sin\theta}\delta(r - r_0)\delta(\theta - \theta_0)\delta(\phi - \phi_0). \tag{2.41b}$$

As usual, (r_0, θ_0, ϕ_0) appearing here are the spherical polar coordinates of the point (x_0, y_0, z_0). From (2.41b) one concludes that the transformation from the Cartesian coordinates (x, y, z) to the polar coordinates (r, θ, ϕ) yields

$$(x, y, z) \to (r, \theta, \phi), \qquad dx\, dy\, dz \to r^2 \sin\theta dr d\theta d\phi$$

and

$$\delta(x - x_0, y - y_0, z - z_0) \to \frac{1}{r^2 \sin\theta} \delta(r - r_0)\delta(\theta - \theta_0)\delta(\phi - \phi_0). \quad (2.41c)$$

EXAMPLE 3

Consider finally the distribution $\delta(S) = \delta(x^2 - a^2)$, which is *assumed* to be concentrated on the planes $x = a$ and $x = -a$. Here $a > 0$ stands for a given constant. In accordance with the *assumption*, we guess that

$$\delta(x^2 - a^2) = A\delta(x - a) + B\delta(x + a), \quad (2.42a)$$

where A and B stand for constants to be determined appropriately. Note that we know the meanings of $\delta(x - a)$ and $\delta(x + a)$ but not $\delta(x^2 - a^2)$. Hence, to compute the above-mentioned numbers A and B and, consequently, to give a meaning to $\delta(x^2 - a^2)$, let us write formally

$$< \delta(x^2 - a^2), \varphi(x) > = \int_{-\infty}^{\infty} \delta(x^2 - a^2)\varphi(x)\, dx \quad (2.42b)$$

and make a change of variable by $x^2 - a^2 = \xi$ to display $\delta(\xi)$. It is obvious that this transformation is not one-to-one, and the inverse transformation has different expressions in the regions $x > 0$ and $x < 0$, namely (see Fig. 2.7),

$$x = \begin{cases} \sqrt{\xi + a^2} & \text{when } x > 0 \\ -\sqrt{\xi + a^2} & \text{when } x < 0. \end{cases}$$

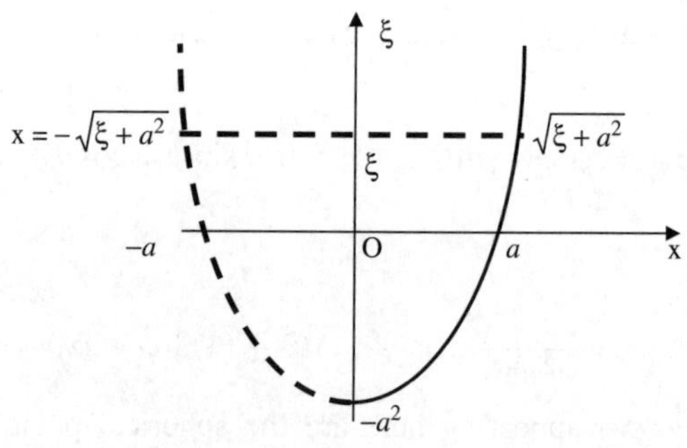

Figure 2.7. One-to-one branches of the transformation $x^2 - a^2 = \xi$.

To avoid the difficulties coming from this multivaluedness, let us divide the support of the distribution $\delta(x^2 - a^2)$ into two parts such that on each of them the transformation in question is one-to-one. Thus (2.42b) is replaced by

$$
\begin{aligned}
&< \delta(x^2 - a^2), \varphi(x) > \\
&= \int_{-\infty}^{0} \delta(x^2 - a^2)\varphi(x)dx + \int_{0}^{\infty} \delta(x^2 - a^2)\varphi(x)dx. \\
&= -\int_{\infty}^{-a^2} \delta(\xi)\varphi(-\sqrt{\xi + a^2})\frac{d\xi}{2\sqrt{\xi + a^2}} \\
&\quad + \int_{-a^2}^{\infty} \delta(\xi)\varphi(\sqrt{\xi + a^2})\frac{d\xi}{2\sqrt{\xi + a^2}} \\
&= \frac{1}{2a}\varphi(-a) + \frac{1}{2a}\varphi(a) \\
&=< \frac{1}{2a}\delta(x - a), \varphi(x) > + < \frac{1}{2a}\delta(x + a), \varphi(x) > \\
&=< \frac{1}{2a}\delta(x - a) + \frac{1}{2a}\delta(x + a), \varphi(x) > .
\end{aligned}
$$

A comparison of this latter with (2.42a) shows that $A = B = 1/(2a)$ and*

$$\delta(x^2 - a^2) = \frac{1}{2a}[\delta(x - a) + \delta(x + a)]. \tag{2.42c}$$

PROBLEMS

1. Let $a \in R$ and $a \neq 0$. Then show that

$$\lim_{a \to 0} \left\{ \frac{1}{|a|} e^{-ixy/a} \right\} = 2\pi\delta(x)\delta(y). \tag{2.43}$$

2. By a formal computation, show that if $\alpha < \beta < \gamma$, then one has

$$\delta((x - \alpha)\delta(x - \beta)\delta(x - \gamma)) = \frac{\delta(x - \alpha)}{|f'(\alpha)|} + \frac{\delta(x - \beta)}{|f'(\beta)|} + \frac{\delta(x - \gamma)}{|f'(\gamma)|},$$

where $f(x) = (x - \alpha)(x - \beta)(x - \gamma)$. Note that $f'(\alpha) > 0$, $f'(\gamma) > 0$ while $f'(\beta) < 0$. Generalize the result to the case of $\delta(f(x))$, where $f(x)$ is a polynomial of degree n.

*This expression was first written and used by Dirac as the definition of $\delta(x^2 - a^2)$ (see Dirac [7, p. 60]).

2.8.2 Vector-Valued Distributions

If the Cartesian components of a vector **A**, say A_1, A_2, A_3, are distributions, then **A** is referred to as a *distribution-valued vector* or a *vector-valued distribution*. The product of such a vector with a constant or a scalar function $\psi(x,y,z)$ having the properties mentioned in Section 2.5.1 are as follows:

$$\psi \mathbf{A} = (\psi A_1)\mathbf{e}_1 + (\psi A_2)\mathbf{e}_2 + (\psi A_3)\mathbf{e}_3. \tag{2.44a}$$

When the components of a vector $\boldsymbol{a}$ are same as the aforementioned function $\psi(x,y,z)$, the scalar (i.e., dot) and vector products of $\boldsymbol{a}$ and **A** are defined by straightforward extensions of the usual definitions, namely,

$$\begin{aligned} < \mathbf{A}\cdot\boldsymbol{a}, \varphi > &= < A_1 a_1 + A_2 a_2 + A_3 a_3, \varphi > \\ &= < A_1, a_1\varphi > + < A_2, a_2\varphi > + < A_3, a_3\varphi > \end{aligned} \tag{2.44b}$$

and

$$\begin{aligned} < \mathbf{A}\times\boldsymbol{a}, \varphi > &= < A_2 a_3 - A_3 a_2, \varphi > \mathbf{e}_1 \\ &\qquad + < A_3 a_1 - A_1 a_3, \varphi > \mathbf{e}_2 + \ldots \\ &= [< A_2, a_3\varphi > - < A_3, a_2\varphi >]\mathbf{e}_1 \\ &\qquad + [< A_3, a_1\varphi > - < A_1, a_3\varphi >]\mathbf{e}_2 + \ldots . \end{aligned} \tag{2.44c}$$

Similarly, the straightforward extensions of the divergence and curl of a distribution vector as well as the gradient of a scalar distribution are as follows:

$$\begin{aligned} < \text{div}\mathbf{A}, \varphi > &= < A_1, -\partial\varphi/\partial x > + < A_2, -\partial\varphi/\partial y > \\ &\qquad + < A_3, -\partial\varphi/\partial z > \end{aligned} \tag{2.45a}$$

$$\begin{aligned} < \text{curl}\mathbf{A}, \varphi > &= [< A_3, -\partial\varphi/\partial y > - < A_2, -\partial\varphi/\partial z >]\,\mathbf{e}_1 \\ &\qquad + [< A_1, -\partial\varphi/\partial z > - < A_3, -\partial\varphi/\partial x >]\,\mathbf{e}_2 + \ldots \end{aligned} \tag{2.45b}$$

$$\begin{aligned} < \text{grad}f, \varphi > &= < f, -\partial\varphi/\partial x > \mathbf{e}_1 + < f, -\partial\varphi/\partial y > \mathbf{e}_2 \\ &\qquad + < f, -\partial\varphi/\partial z > \mathbf{e}_3. \end{aligned} \tag{2.45c}$$

Now we are ready to state the following.

THEOREM 2.5

Let **A** be a vector field defined outside a regular surface S such that:

i. It is bounded at every point except perhaps at points at infinity.

ii. It has continuous first-order derivatives everywhere except perhaps at points at infinity.

Then one has

$$\operatorname{div}\mathbf{A} = \{\operatorname{div}\mathbf{A}\} + [[\mathbf{n} \cdot \mathbf{A}]]\delta(S), \tag{2.46a}$$

$$\operatorname{curl}\mathbf{A} = \{\operatorname{curl}\mathbf{A}\} + [[\mathbf{n} \times \mathbf{A}]]\delta(S), \tag{2.46b}$$

$$\operatorname{div}[\operatorname{curl}\mathbf{A}] \equiv 0. \tag{2.46c}$$

Here **n** stands for the arbitrarily oriented unit normal vector to S while the quantities given by the double brackets $[[\cdot]]$ are the increments of the quantities in question when one crosses S in the direction shown by **n**.

THEOREM 2.6

Let $f(x, y, z)$ be defined outside a regular surface S such that it is bounded and has continuous first-order derivatives everywhere except perhaps at points at infinity. Then one has

$$\operatorname{grad}f = \{\operatorname{grad}f\} + [[f\mathbf{n}]]\delta(S), \tag{2.47a}$$

$$\operatorname{curl}[\operatorname{grad}f] \equiv 0. \tag{2.47b}$$

Here **n** stands for the arbitrarily oriented unit normal vector to S while the quantity given by the double bracket $[[\cdot]]$ is the increment of the quantity in question when one passes across S in the direction shown by **n**.

Proof [5]

Let S be composed of certain closed or open regular surfaces $S_1, S_2, \ldots, S_n$ (see Fig. 2.8). It is worthwhile to remark that the increments $[[\mathbf{n} \cdot \mathbf{A}]], [[\mathbf{n} \times \mathbf{A}]]$, and $[[f\ \mathbf{n}]]$, which depend only on the coordinates on S, say ξ and η, play roles of constant parameters in regard to the distribution $\delta(S)$ which depends on the third coordinate, say ζ. Hence the vectors $[[\mathbf{n} \cdot \mathbf{A}]]\delta(S), [[\mathbf{n} \times \mathbf{A}]]\delta(S)$, and $[[\mathrm{f}\mathbf{n}]]\delta(S)$, which appear in (2.46 a,b) and (2.47a), satisfy all requirements mentioned in Section 2.5.1.

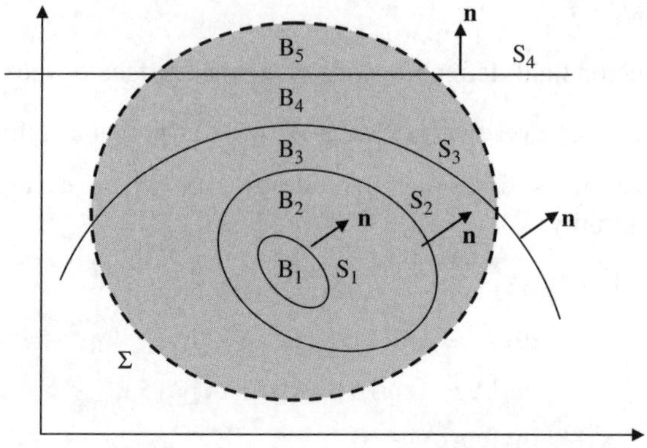

Figure 2.8. Space divided by regular surfaces.

To prove the formulas (2.46a)–(2.47b), consider any test function $\varphi \in D$ and a sphere Σ such that it involves the support of φ as well as all the closed surfaces $S_j \subset S$, if any. Let the regions bounded by Σ and S_j be named as B_1, B_2, and so on (see Fig. 2.8). In this case, starting from (2.45a) one writes successively:

$$
\begin{aligned}
< \operatorname{div}\mathbf{A}, \varphi > &=< A_1, -\partial\varphi/\partial x > + \ldots. \\
&= -\iiint \{\mathbf{A}\cdot\operatorname{grad}\varphi\}\, dxdydz \\
&= -\int_{\cup B_j} \mathbf{A}\cdot\operatorname{grad}\varphi dv, \qquad (dv = dxdydz) \\
&= \int_{\cup B_j} \{\varphi\operatorname{div}\mathbf{A} - \operatorname{div}(\varphi\mathbf{A})\} dv \\
&= \iiint \{\varphi\operatorname{div}\mathbf{A}\} dv - \int_{\cup\partial B_j} \varphi A_n dS.
\end{aligned}
$$

Note that these equalities are based on the classical formula [2, p. 297]

$$\operatorname{div}(\varphi\mathbf{A}) = \varphi\operatorname{div}\mathbf{A} + \mathbf{A}\cdot\operatorname{grad}\varphi$$

of the vector analysis as well as the Gauss–Ostrogradski formula (divergence theorem). A_n appearing in the last term means, for each ∂B_j, the projection of $\mathbf{A}$ on the normal vector directed outwards the region B_j. By virtue of the fact that $\varphi \equiv 0$ on and outside Σ, and the normal vectors on ∂B_j pertinent to adjacent regions are inverse to each other, the last line of the above formulas reduces to

$$< \mathrm{div}\mathbf{A},\varphi > = \iiint \{\varphi \mathrm{div}\mathbf{A}\}\, dv + \int_S \varphi[[\mathbf{n}\cdot\mathbf{A}]]dS$$
$$=< \{\mathrm{div}\mathbf{A}\} + [[\mathbf{n}\cdot\mathbf{A}]]\delta(S), \varphi >,$$

which is nothing but (2.46a).

Quite similarly, starting from (2.45b) one writes successively

$$< \mathrm{curl}\mathbf{A},\varphi > = [< A_3, -\partial\varphi/\partial y > - < A_2, -\partial\varphi/\partial z >]\,\mathbf{e}_1 + \ldots$$
$$= -\iiint [\mathrm{grad}\varphi \times \mathbf{A}]dxdydz$$
$$= -\int_{\cup B_j} \{\mathrm{grad}\varphi \times \mathbf{A}\}dv$$
$$= \int_{\cup B_j} \{\varphi \mathrm{curl}\mathbf{A} - \mathrm{curl}(\varphi\mathbf{A})\}\, dv$$
$$= \int_{\cup B_j} \{\varphi \mathrm{curl}\mathbf{A}\}dv - \int_{\cup\partial B_j} \mathbf{n}\times(\varphi\mathbf{A})dS$$
$$= \iiint \varphi\{\mathrm{curl}\mathbf{A}\}dv + \int_S \varphi[[\mathbf{n}\times\mathbf{A}]]\, dS$$
$$=< \{\mathrm{curl}\mathbf{A}\} + [[\mathbf{n}\times\mathbf{A}]]\delta(S), \varphi>,$$

which proves (2.46b). The fifth equality above is based on the known formula [2, p. 296]

$$\int_B \mathrm{curl}\mathbf{F}\,\mathbf{dv} = \int_{\partial B} \mathbf{n}\times\mathbf{F}\,dS$$

of the vector analysis. Here the normal vector on ∂B is directed outwards the region B.

Consider now div[curl**A**] in the sense of distribution. By showing the components of curl**A** by $[\mathrm{curl}\mathbf{A}]_1$, $[\mathrm{curl}\mathbf{A}]_2$ and so on, one gets

$$< \mathrm{div}[\mathrm{curl}\mathbf{A}], \varphi > =< [\mathrm{curl}\mathbf{A}]_1, -\partial\varphi/\partial x > + < [\mathrm{curl}\mathbf{A}]_2, -\partial\varphi/\partial y >$$
$$+ < [\mathrm{curl}\mathbf{A}]_3, -\partial\varphi/\partial z >$$

$$
\begin{aligned}
&= \left\langle A_3, -\frac{\partial}{\partial y}\left(-\frac{\partial \varphi}{\partial x}\right)\right\rangle - \left\langle A_2, -\frac{\partial}{\partial z}\left(-\frac{\partial \varphi}{\partial x}\right)\right\rangle \\
&+ \left\langle A_1, -\frac{\partial}{\partial z}\left(-\frac{\partial \varphi}{\partial y}\right)\right\rangle - \left\langle A_3, -\frac{\partial}{\partial x}\left(-\frac{\partial \varphi}{\partial y}\right)\right\rangle \\
&+ \left\langle A_2, -\frac{\partial}{\partial x}\left(-\frac{\partial \varphi}{\partial z}\right)\right\rangle - \left\langle A_1, -\frac{\partial}{\partial y}\left(-\frac{\partial \varphi}{\partial z}\right)\right\rangle \\
&\equiv 0.
\end{aligned}
$$

This proves (2.46c).

To prove (2.47a), we write successively

$$
\begin{aligned}
< \text{grad} f, \varphi > &= < f, -\partial\varphi/\partial x > \mathbf{e}_1 + \ldots \\
&= -\iiint f \,\text{grad}\varphi \, dxdydz \\
&= -\int_{\cup \mathrm{B_j}} \{\text{grad}(f\varphi) - \varphi \text{grad} f\} dv \\
&= -\int_{\cup \partial \mathrm{B_j}} (\mathrm{f}\mathbf{n})\varphi dS + \int_{\cup \mathrm{B_j}} \{\varphi \text{grad} f\} dv \\
&= \int_{\mathrm{S}} [[\mathrm{f}\mathbf{n}]]\varphi dS + \iiint \{\text{grad} f\}\varphi dv \\
&= < \{\text{grad} f\} + [[f\mathbf{n}]]\delta(S), \varphi > .
\end{aligned}
$$

Notice that in writing the fourth line we have used the identity [2, p. 296]

$$
\int_B \text{grad} f \, dv = \int_{\partial B} (f\mathbf{n}) \, dS
$$

of the vector analysis. By considering the definitions, we write also $< \text{curl}(\text{grad} f), \varphi >$

$$
\begin{aligned}
&= [< [\text{grad} f]_3, -\partial\varphi/\partial y > - < [\text{grad} f]_2, -\partial\varphi/\partial z >]\mathbf{e}_1 + \ldots \\
&= \left[\left\langle f, -\frac{\partial}{\partial z}\left(-\frac{\partial \varphi}{\partial y}\right)\right\rangle - \left\langle f, -\frac{\partial}{\partial y}\left(-\frac{\partial \varphi}{\partial z}\right)\right\rangle\right] \mathbf{e}_1 + \ldots \\
&\equiv 0,
\end{aligned}
$$

which proves (2.47b). ■

PROBLEMS

1. **(a)** Let ρ denote the distance of the point (x, y) to a fixed point (α, β) in the two-dimensional space R^2. Then show that

$$\Delta\left(\log\left\{\frac{1}{\rho}\right\}\right) = -2\pi\,\delta(x-\alpha)\delta(y-\beta).$$

(b) Let r denote the distance of the point (x, y, z) to a fixed point (α, β, γ) in the three dimensional space R^3. Then show that

$$\Delta\left(\frac{1}{r}\right) = -4\pi\,\delta(x-\alpha)\delta(y-\beta)\delta(z-\gamma).$$

2. Show that

$$\Delta\left(\frac{1}{r}e^{ikr}\right) + k^2\left(\frac{1}{r}e^{ikr}\right) = -4\pi\,\delta(x-\alpha)\delta(y-\beta)\delta(z-\gamma),$$

where r denotes the distance of the point (x, y, z) to a fixed point (α, β, γ) while k is any (positive or negative) number.

3. **(a)** Assume that the half-spaces $y < b$ and $y > b$ are filled with dielectric materials of permittivities ε_1 and ε_2, respectively, while a point charge of amount Q is located at the point (a, b, c). Then show that the electrostatic potential $u(x, y, z)$ created by Q, which satisfies the Laplace equation $\Delta u = 0$ everywhere outside the plane $y = b$, satisfies also the following boundary conditions on the interface $y = b$:

$$[[u]] = 0, \qquad \left[\left[\varepsilon(y)\frac{\partial u}{\partial n}\right]\right] = -Q\delta(x-a)\delta(z-c).$$

(b) If the point charge Q is replaced by a point dipole of moment $\mathbf{p} = P\mathbf{e}_x$, then the relations satisfied on the plane $y = b$ become

$$[[u]] = 0, \qquad \left[\left[\varepsilon(y)\frac{\partial u}{\partial n}\right]\right] = Pd'(x-a)\delta(z-c).$$

(c) When the moment of the dipole is $\mathbf{p} = P\mathbf{e}_y$ while $\varepsilon_1 = \varepsilon_2 \equiv \varepsilon$, the boundary conditions on the plane $y = b$ are changed to

$$\left[\left[\frac{\partial u}{\partial n}\right]\right] = 0, \qquad [[u]] = \left(\frac{P}{\varepsilon}\right)\delta(x-a)\delta(z-c).$$

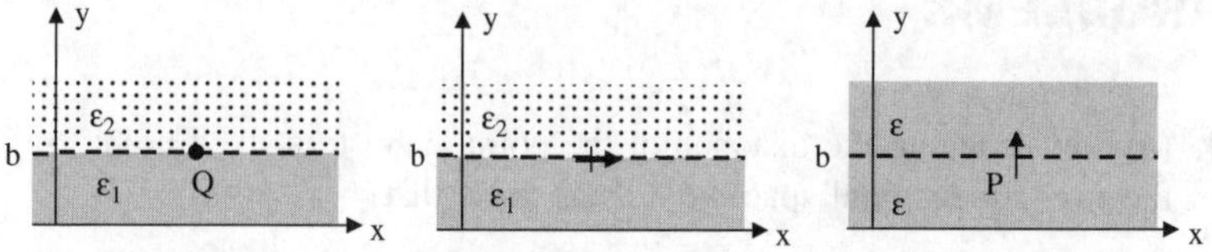

4. **(a)** Let (r, ϕ, z) denote the usual cylindrical polar coordinates while the regions B_1 and B_2 are defined as

$$B_1 = \{(r, \phi, z) : r \in (0, \infty), \phi \in (0, \psi), z \in (-\infty, \infty)\},$$
$$B_2 = \{(r, \phi, z) : r \in (0, \infty), \phi \in (\psi, \alpha), z \in (-\infty, \infty)\},$$

where $\alpha \in (0, 2\pi)$ and $\psi \in (0, \alpha)$ stand for two given angles. Assume that the regions B_1 and B_2 are filled with dielectric materials of permittivities ε_1 and ε_2, respectively, while a point charge of amount Q is located at the point (a, ψ, c). Then show that the electrostatic potential $u(r, \phi, z)$ created by Q, which satisfies in the regions B_1 and B_2 the Laplace equation $\Delta u = 0$, satisfies on the interface $\phi = \psi$ the following boundary conditions:

$$[[u]] = 0, \qquad \left[\left[\varepsilon(\phi)\frac{\partial u}{\partial n}\right]\right] = -Q\delta(r-a)\delta(z-c).$$

(b) If the point charge Q is replaced by a point dipole of moment $\mathbf{p} = P\mathbf{e}_r$, then the relations satisfied on the interface $\phi = \psi$ become

$$[[u]] = 0, \qquad \left[\left[\varepsilon(\phi)\frac{\partial u}{\partial n}\right]\right] = P\frac{r}{a}\delta'(r-a)\delta(z-c).$$

(c) When the moment of the dipole is $\mathbf{p} = P\mathbf{e}_\phi$ while $\varepsilon_1 = \varepsilon_2 \equiv \varepsilon$, the boundary conditions on the interface $\phi = \psi$ are changed to

$$\left[\left[\frac{\partial u}{\partial n}\right]\right] = 0, \qquad [[u]] = \left(\frac{P}{\varepsilon}\right)\delta(r-a)\delta(z-c).$$

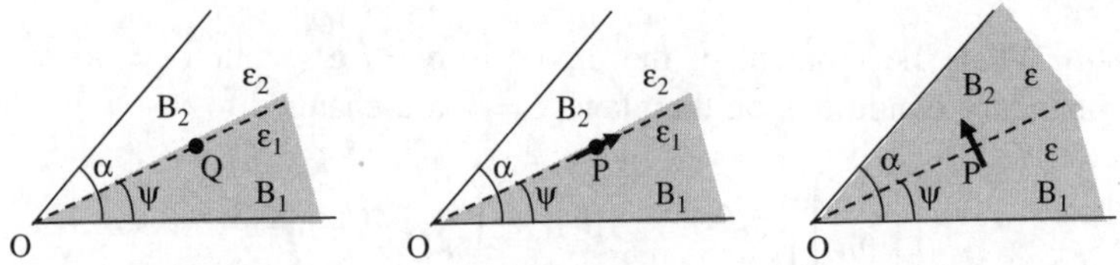

5. **(a)** Let (r, ϕ, z) denote the usual cylindrical polar coordinates while the regions B_1 and B_2 are defined as

$$B_1 = \{(r, \phi, z) : r \in (0, a), \phi \in (0, \alpha), z \in (-\infty, \infty)\},$$
$$B_2 = \{(r, \phi, z) : r \in (a, \infty), \phi \in (0, \alpha), z \in (-\infty, \infty)\},$$

where $\alpha \in (0, 2\pi)$ and $a \in (0, \infty)$ stand for two given constants. Assume that the regions B_1 and B_2 are filled with dielectric materials of permittivities ε_1 and ε_2, respectively, while a point charge of amount Q is located at the point (a, ψ, c). Then show that the electrostatic potential $u(r, \phi, z)$ created by Q, which satisfies in the regions B_1 and B_2 the Laplace equation $\Delta u = 0$, satisfies on the interface $r = a$ the following boundary conditions:

$$[[u]] = 0, \qquad \left[\left[\varepsilon(r)\frac{\partial u}{\partial n}\right]\right] = -\left(\frac{Q}{a}\right)\delta(\phi - \psi)\delta(z - c).$$

(b) If the point charge Q is replaced by a point dipole of moment $\mathbf{p} = P\mathbf{e}_\phi$, then the relations satisfied on the interface $r = a$ become

$$[[u]] = 0, \qquad \left[\left[\varepsilon(r)\frac{\partial u}{\partial n}\right]\right] = \left(\frac{P}{a^2}\right)\delta'(\phi - \psi)\delta(z - c).$$

(c) When the moment of the dipole is $\mathbf{p} = P\mathbf{e}_r$ while $\varepsilon_1 = \varepsilon_2 \equiv \varepsilon$, the boundary conditions on the interface $r = a$ are changed to

$$\left[\left[\frac{\partial u}{\partial n}\right]\right] + \frac{1}{a}[[u]] = 0, \qquad [[u]] = P/(a\varepsilon)\delta(\phi - \psi)\delta(z - c).$$

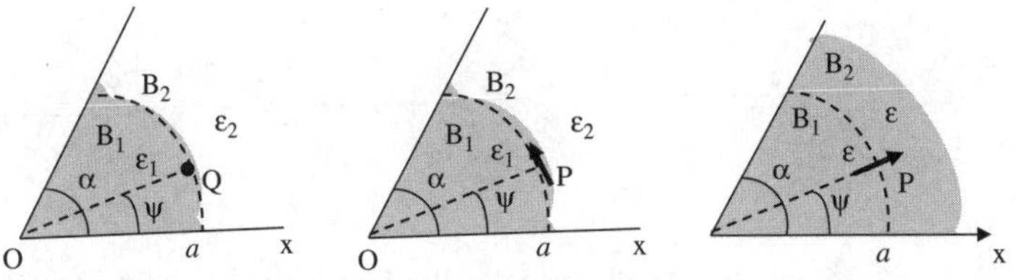

6. **(a)** Let (r, θ, ϕ) denote the usual spherical polar coordinates while the regions B_1 and B_2 are defined as

$$B_1 = \{(r, \theta, \phi) : r \in [0, a), \theta \in [0, \pi], \phi \in [0, 2\pi)\},$$
$$B_2 = \{(r, \theta, \phi) : r \in [a, \infty), \theta \in [0, \pi], \phi \in [0, 2\pi)\},$$

where $a \in (0, \infty)$ stands for a given constant. Assume that the regions B_1 and B_2 are filled with dielectric materials of permittivities ε_1 and ε_2, respectively, while a point charge of amount Q is located at the point

(a, ψ, α). Then show that the electrostatic potential u $(\mathrm{r}, \theta, \phi)$ created by Q, which satisfies in the regions B_1 and B_2 the Laplace equation $\Delta \mathrm{u} = 0$, satisfies on the interface $r = a$ the following boundary conditions:

$$[[u]] = 0, \qquad \left[\left[\varepsilon(\mathrm{r})\frac{\partial \mathrm{u}}{\partial \mathrm{n}}\right]\right] = -Q/(a^2 \sin\psi)\delta(\theta - \psi)\delta(\phi - \alpha).$$

(b) If the point charge Q is replaced by a point dipole of moment $\mathbf{p} = P\mathbf{e}_\theta$, then the relations satisfied on the interface $\mathrm{r} = a$ become

$$[[u]] = 0, \qquad \left[\left[\varepsilon(r)\frac{\partial u}{\partial n}\right]\right] = P/(a^3 \sin\psi)\, \delta'(\theta - \psi)\delta(\phi - \alpha).$$

(c) When the moment of the dipole is $\mathbf{p} = \mathrm{P}\mathbf{e}_\phi$, then the boundary conditions on the interface $r = a$ are

$$[[u]] = 0, \qquad \left[\left[\varepsilon(r)\frac{\partial u}{\partial n}\right]\right] = P/(a^3 \sin^2\psi)\delta(\theta - \psi)\, \delta'(\phi - \alpha).$$

(d) If the moment of the dipole is $\mathbf{p} = P\mathbf{e}_r$ while $\varepsilon_1 = \varepsilon_2 \equiv \varepsilon$, then the boundary conditions on the interface $r = a$ are changed to

$$\left[\left[\frac{\partial u}{\partial n}\right]\right] + \frac{2}{a}[[u]] = 0, \qquad [[u]] = P/(\varepsilon a^2 \sin\psi)\delta(\theta - \psi)\delta(\phi - \alpha)$$

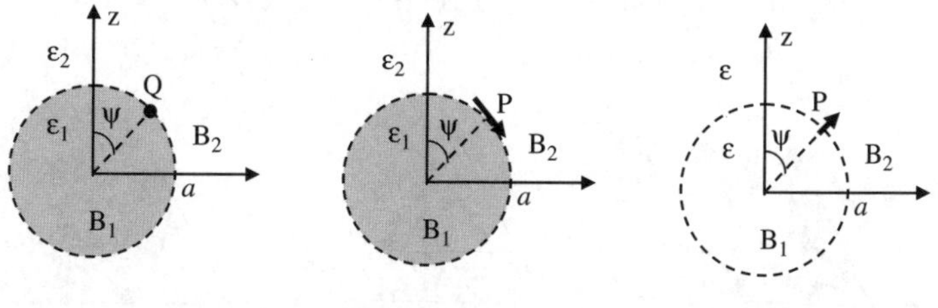

7. **(a)** Let (r, θ, ϕ) denote the usual spherical polar coordinates while the regions B_1 and B_2 are defined as

$$B_1 = \{(r, \theta, \phi) : r \in [0, \infty), \theta \in [0, \psi), \phi \in [0, 2\pi)\},$$
$$B_2 = \{(r, \theta, \phi) : r \in [0, \infty), \theta \in (\psi, \pi], \phi \in [0, 2\pi)\},$$

where $\psi \in (0, \pi)$ stands for a given angle. Assume that the regions B_1 and B_2 are filled with dielectric materials of permittivities ε_1 and ε_2, respectively, while a point charge of amount Q is located at the point (a, ψ, α). Then show that the electrostatic potential $u(r, \theta, \phi)$ created by

Q, which satisfies in the regions B_1 and B_2 the Laplace equation $\Delta u = 0$, satisfies on interface cone $\theta = \psi$ the following boundary conditions:

$$[[u]] = 0, \qquad \left[\left[\varepsilon(\theta)\frac{\partial u}{\partial n}\right]\right] = -Q/(a \sin\psi)\delta(r-a)\delta(\phi-\alpha).$$

(b) If the point charge Q is replaced by a point dipol of moment $\mathbf{p} = P\mathbf{e}_r$, then the relations satisfied on the interface $\theta = \psi$ become

$$[[u]] = 0, \qquad \left[\left[\varepsilon(\theta)\frac{\partial u}{\partial n}\right]\right] = P/(a^2 \sin\psi) r\delta'(\phi-\psi)\delta(\phi-\alpha).$$

(c) When the moment of the dipole is $\mathbf{p} = P\mathbf{e}_\phi$, then the boundary conditions on the inteface $\theta = \psi$ become

$$[[u]] = 0, \qquad \left[\left[\varepsilon(\theta)\frac{\partial u}{\partial n}\right]\right] = P/(a^2 \sin^2\psi)\delta(r-a)\delta'(\phi-\alpha).$$

(d) If the moment of the dipole is $\mathbf{p} = P\mathbf{e}_\theta$ while $\varepsilon_1 = \varepsilon_2 \equiv \varepsilon$, then the boundary conditions on the interface $\theta = \psi$ are changed to

$$\left[\left[\frac{\partial u}{\partial n}\right]\right] + \frac{1}{r}[[u]]\cot\psi = 0, \qquad [[u]] = P/(\varepsilon a \sin\psi)\delta(r-a)\delta(\phi-\alpha).$$

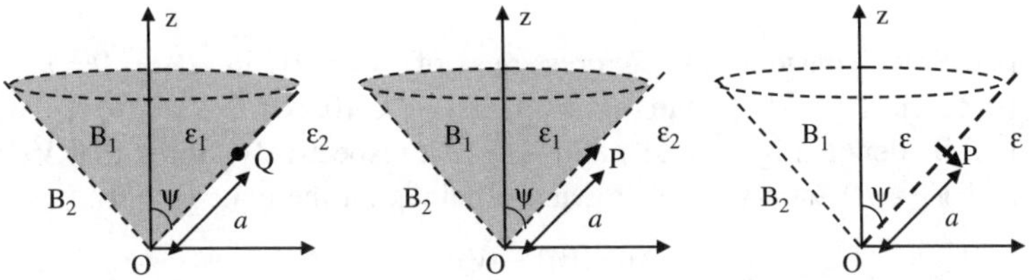

8. Let $V(r,\theta)$ be the potential function created by a charge of amount Q, uniformly distributed on a circle located in the region bounded by two conducting cones (see the accompanying figure). If the cones are earthed, then $V(r,\theta)$ satisfies the equation

$$\Delta V = -\frac{Q}{2\pi\varepsilon c^2 \sin\theta_0}\delta(r-c)\delta(\theta-\theta_0), \qquad r \in (0,\infty), \qquad \theta \in (\theta_1,\theta_2)$$

under the following boundary and edge conditions :

$$V(r,\theta_1) = 0, \qquad V(r,\theta_2) = 0,$$
$$V(r,\theta) \to o(1/r) \qquad \text{as } r \to \infty,$$
$$|V(r,\theta)| < \infty \qquad \text{as } r \to 0.$$

Here $\varepsilon > 0$ stands for the permittivity of the space.

(a) Let the expressions of $V(r,\theta)$ in $B_1 = \{(r,\theta) : r \in (0,c), \theta \in (\theta_1,\theta_2)\}$ and $B_2 = \{(r,\theta) : r \in (c,\infty), \theta \in (\theta_1,\theta_2)\}$ be $V_1(r,\theta)$ and $V_2(r,\theta)$, respectively. Then show that $V_{1,2}(r,\theta)$ satisfy the relations

$$V_2(c,\theta) - V_1(c,\theta) = 0$$

and

$$\frac{\partial}{\partial r}V_2(c,\theta) - \frac{\partial}{\partial r}V_1(c,\theta) = -\frac{Q}{2\pi\varepsilon c^2 \sin\theta_0}\delta(\theta-\theta_0)$$

on the sphere $r = c$. Find the expressions of $V_{1,2}(r,\theta)$ in the form of *series* of Legendre functions.

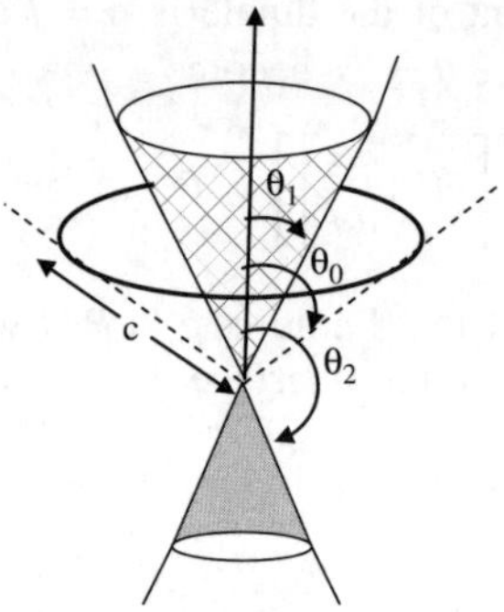

(b) Now consider the expressions of $V(r,\theta)$ in $B_3 = \{(r,\theta) : r \in (0,\infty), \theta \in (\theta_1,\theta_0)\}$ and $B_4 = \{(r,\theta) : r \in (0,\infty), \theta \in (\theta_0,\theta_2)\}$, which will be denoted by $V_3(r,\theta)$ and $V_4(r,\theta)$, respectively. Show that $V_3(r,\theta)$ and $V_4(r,\theta)$ satisfy the following relations on the cone $\theta = \theta_0$:

$$V_4(r,\theta_0) - V_3(r,\theta_0) = 0$$

and

$$\frac{\partial}{\partial\theta}V_2(c,\theta) - \frac{\partial}{\partial\theta}V_1(c,\theta) = -\frac{Q}{2\pi\varepsilon c^2 \sin\theta_0}\delta(r-c).$$

Find the expressions of $V_{3,4}(r,\theta)$ in the following form (Mellin transform):

$$V_j(r,\theta) = \int_L [D_j(\nu)P_{-\nu}(\cos\theta) + E_j(\nu)P_{-\nu}(-\cos\theta)]r^{-\nu}d\nu, \qquad j = 3,4,$$

Here $P_\alpha(\tau)$ stands for the Legendre function bounded at $\tau = 1$. What is the integration path L?

(c) Compare the expressions found in (a) and (b) in the regions $B_1 \cap B_3, B_1 \cap B_4, B_2 \cap B_3$ and $B_2 \cap B_4$.

CHAPTER 3

Maxwell Equations in the Sense of Distribution

3.1 MAXWELL EQUATIONS REDUCED INTO THE VACUUM

The basic equations satisfied by the components of the electromagnetic field, which are usually denoted today by **E, D, B**, and **H**, had been unified in 1873 by Maxwell* as (1.1a–d) [13]. The first two of them are based on rather simple phenomena which were first observed experimentally and then stated theoretically as laws by Ampère† in 1820 [14] and Faraday‡ in 1831 [15], respectively. The third one consists of the differential form of the so-called Gauss§ law, which relates the dielectric displacement to the electric charge density. Finally the last one, which is the basic equation satisfied by the magnetic induction field, by comparing with the third equation, is interpreted as the *nonexistence of the free magnetic charge*. In what follows we will sometimes write these equations by reducing them into the vacuum, namely,

$$\mathrm{curl}\mathbf{H} - \varepsilon_0 \frac{\partial}{\partial t}\mathbf{E} = \mathbf{J} + \frac{\partial}{\partial t}\mathbf{P}, \qquad \mathrm{curl}\mathbf{E} + \mu_0 \frac{\partial}{\partial t}\mathbf{H} = -\frac{\partial}{\partial t}\mathbf{M}, \quad (3.1\mathrm{a,b})$$

$$\varepsilon_0\, \mathrm{div}\mathbf{E} = \rho - \mathrm{div}\mathbf{P}, \qquad \mu_0\, \mathrm{div}\mathbf{H} = -\mathrm{div}\mathbf{M}. \quad (3.1\mathrm{c,d})$$

*J. C. Maxwell (Edinburgh 1831–Cambridge 1879)
†A. M. Ampère (Lyon 1775–Marseilles 1836)
‡M. Faraday (London 1791–Hampton Court 1867)
§C. F. Gauss (Braunschweig 1777–Göttingen 1855)

Discontinuities in the Electromagnetic Field, First Edition. By M. Mithat Idemen.

Published 2011 by John Wiley & Sons, Inc.

Here the fields **P** and **M** appearing in (3.1a–d) represent the volume densities of the *electric polarization* and *magnetic polarization*, respectively, which occur inside the material filling the space. In the vacuum one has $\mathbf{P} \equiv 0, \mathbf{M} \equiv 0$.

Inside the conducting materials the current densities are thought of to be composed of two terms, namely,

$$\mathbf{J} = \mathbf{J}_v + \sigma\mathbf{E}. \tag{3.2}$$

Here $\mathbf{J}_v$ represents the contributions of free charges that are in motion (*convective* current) while the other (namely, $\sigma\mathbf{E}$) is due to the conductivity of the material (*conduction* current). So the coefficient σ is referred to as the *electric conductivity* of the material. Although (3.2) is written as if it is linear with respect to **E**, in general the *constitutive parameter* σ may depend also on **E** and **H** (the case of nonlinear materials) as well as some various other parameters such as the frequency, the position of the observation point, the time, the temperature, and so on. From (3.1a–d) one gets the following relation (*equation of continuity*) between the current and charge densities:

$$\mathrm{div}\mathbf{J} + \partial\rho/\partial t = 0. \tag{3.3}$$

This shows that the current and charge densities cannot be thought independently.

As is well known, the fields **D** and **B** appearing in the equations (1.1a–d) are defined through

$$\mathbf{D} = \varepsilon_0\mathbf{E} + \mathbf{P} \equiv \varepsilon\mathbf{E}, \qquad \mathbf{B} = \mu_0\mathbf{H} + \mathbf{M} \equiv \mu\mathbf{H}. \tag{3.4a,b}$$

Here ε and μ are referred to as the dielectric permittivity and magnetic permeability of the material filling the region in question. They change, in general, with the field components as well as with various other parameters as stated above in connection with σ. Unless the contrary is stated clearly, in what follows we will assume that the space is composed of finitely many regions such that in each of them the constitutive parameters ε, μ, and σ are constant scalars. A material whose constitutive parameters are constant scalars is referred to as a *simple material*. On the interfaces between regions filled with different simple materials, along with the constitutive parameters ε, μ, and σ, the field components **E**, **D**, **B**, and **H** also display jump discontinuities. Therefore, in order to reveal the behavior of an electromagnetic field created by certain (known) sources in a space composed of various regions having different physical properties, in addition to the basic equations (3.1a)–(3.2), one needs to know also the jump discontinuities on the interfaces between different regions. The expressions of these discontinuities are called the *boundary conditions*. Different types of boundary conditions will lead different solutions to an electromagnetic problem. But the physical phenomenon to be observed experimentally is unique. That means that the boundary conditions can (and must) be completely known whenever the physical structure of the space is known. So the determination of the boundary conditions constitute a crucial problem in connection with the propagation and scattering of electromagnetic waves. One of the three problems,* which are the main subjects of this monograph, is this topic.

We believe (and claim) that the boundary conditions in question cannot be stated independently from the Maxwell equations; they can (and must) be derived only from (3.1a–d). In this struggle our unique and solid support will be the following assumption.

Assumption. The Maxwell equations (3.1a–d) are valid in the whole of the four-dimensional space in the sense of distribution (see Idemen [5, 16]).

As we have already remarked in the introduction, this assumption not only legitimizes the use of Dirac distributions in the expressions of the source densities ρ and $\mathbf{J}$, but also claims that the field components $\mathbf{E}$, $\mathbf{H}$, $\mathbf{B}$, and $\mathbf{M}$ themselves are distributions. Let the union of the surfaces, which carry surface or line or point charges and currents as well as those on which there are jump discontinuities, be denoted by S. Then the singular parts of the distributions which will take place in the expressions of different terms appearing in the Maxwell equations will be concentrated only on S. Suppose now that S is divided into finitely many parts, say S_j $(j = 1, 2, \ldots, n)$ such that the equation of S_j with respect to a certain reference coordinate system $O(x, y, z)$ is known as

$$w_j(x, y, z, t) = 0, \tag{3.5}$$

and there is a certain orthogonal coordinate system (u, v, w_j) such that the transformation from (x, y, z) to (u, v, w_j) is one-to-one (see Section 2.8.1, Example 3). In this case, in accordance with the

*These problems are the boundary, edge, and tip conditions.

Theorem 2.4 of Section 2.7, the singular parts of the distributions concentrated on S_j (i.e., on the surface $w_j = 0$) will be of the form

$$f = c_0\delta(w_j) + c_1\delta'(w_j) + \cdots + c_N\delta^{(N)}(w_j).$$

Without loss of generality, we can assume that the number N is same for all terms (the coefficients of nonexisting terms are equal to naught). Thus by omitting N as well as the index j, we write

$$\rho = \{\rho\} + \sum_{k=0} \rho_k\delta^{(k)}(w) \tag{3.6a}$$

$$\mathbf{E} = \{\mathbf{E}\} + \sum_{k=0} \mathbf{E}_k\delta^{(k)}(w) \tag{3.6b}$$

(similar expressions for other quantities)

Here the coefficients $\rho_k, \mathbf{E}_k$, and so on, stand for certain functions which are defined on S and depend on the coordinates (u, v) and time t.

It is obvious that there is a close connection between $\delta(w)$ appearing in (3.6a,b, ...) and $\delta(S)$ taking place in (2.46a,b)–(2.47a). To reveal this connection, let us compute the gradient of a function $f(x, y, z, t)$, which has a jump discontinuity on S, in the sense of distribution and compare the result with (2.47a). Since on each part of S (i.e., on S_j) the transformation $(x, y, z) \leftrightarrow (u, v, w)$ is one-to-one, in one region limited by S we have $w > 0$ while in the other we have $w < 0$ (see Fig. 3.1). Without loss of generality, we can assume that the unit normal vector to S, say $\mathbf{n}$, is directed into the region of $w > 0$. Thus we write first

$$f(x,y,z,t) = \begin{cases} f_1(x,y,z,t), & \text{in the region of } w < 0 \\ f_2(x,y,z,t), & \text{in the region of } w > 0 \end{cases}$$
$$= f_2H(w) + f_1[1 - H(w)] \tag{3.7}$$

and then,

$$\begin{aligned} \text{grad}f &= \text{grad}\{f\} \\ &= \text{H}(w)\text{grad}f_2 + [1 - H(w)]\text{grad}f_1 \\ &\quad + f_2\text{grad}H(w) - f_1\text{grad}H(w) \\ &= \{\text{grad}f\} + [f_2 - f_1]\frac{dH(w)}{dw}\text{grad}w \\ &= \{\text{grad}f\} + [[f]]\delta(w)\text{grad}w \\ &= \{\text{grad}f\} + [[f]]\delta(w)|\text{grad}w|\mathbf{n}. \end{aligned} \tag{3.8a}$$

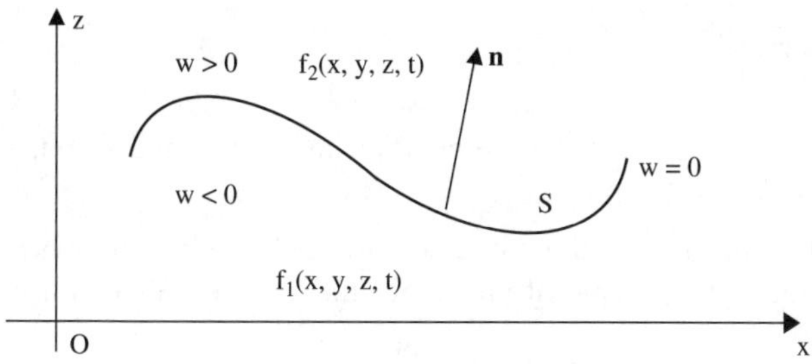

Figure 3.1. A function having a bounded jump discontinuity on S.

A comparison of (3.8a) with (2.47a) shows that

$$\delta(S) = |\text{grad}w|\delta(w). \tag{3.8b}$$

3.1.1 Some Simple Examples

1. When S consists of the sphere of radius a, one has $w = r - a$, which yields grad$w = \mathbf{r}/r = \mathbf{e}_r = \mathbf{n}$ and $\delta(S) = \delta(r - a)$. For an oscillating sphere defined through $r = a(1 + \varepsilon \sin t)$, where t stands for the time, one has $w = r - a(1 + \varepsilon \sin t)$ and $\delta(S) = \delta(r - a - a\varepsilon \sin t)$.

2. If S is a circular cylinder with radius a, then one writes $w = (\rho - a)$ and grad$w = \boldsymbol{\rho}/\rho = \mathbf{e}_\rho = \mathbf{n}$, which gives $\delta(S) = \delta(\rho - a)$. Here ρ stands for the distance from the axis of the cylinder.

3. In the case when $S = S_1 \cup S_2$ consists of the boundary of a wedge (see Fig. 3.2a), one has $w = \phi - \phi_0$ on S_2 and $w = \phi + \phi_0$ on S_1, which yield grad$w = (1/\rho)\mathbf{e}_\phi$ on both S_1 and S_2. Hence $\delta(S) = (1/\rho)\delta(\phi \pm \phi_0)$.

4. If S is a rotational cone with apex angle equal to θ_0, then one has $w = \theta - \theta_0$, grad$w = (1/r)\mathbf{e}_\theta$, and $\delta(S) = (1/r)\delta(\theta - \theta_0)$. Here r denotes the radial distance from the origin (see Fig. 3.2b).

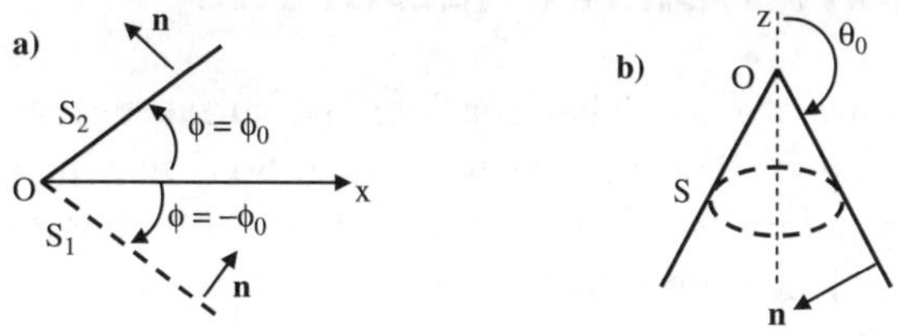

Figure 3.2. Discontinuity surfaces mentioned in Examples 3 and 4.

PROBLEMS

1. Discuss the dimensions of the coefficients $\rho_k, \mathbf{E}_k$, and so on, which take place in (3.6a,b, ...).

2. Show that in the vacuum any electromagnetic field can always be represented through a couple of potential functions $(V, \mathbf{A})$ as follows:

$$\mathbf{B} = \mathrm{curl}\mathbf{A}, \qquad E = -\partial\mathbf{A}/\partial t - \mathrm{grad}V, \qquad \mathbf{H} = \mathbf{B}/\mu_0, \qquad \mathbf{D} = \varepsilon_0\mathbf{E}.$$

Here V and $\mathbf{A}$ satisfy the following wave equations:

$$\Delta V - \varepsilon_0\mu_0\frac{\partial^2}{\partial t^2}V = -\frac{\rho}{\varepsilon_0},$$

$$\Delta\mathbf{A} - \varepsilon_0\mu_0\frac{\partial^2}{\partial t^2}\mathbf{A} = -\mu_0\mathbf{J}_\nu,$$

$$\mathrm{div}\mathbf{A} + \varepsilon_0\mu_0\frac{\partial}{\partial \mathrm{t}}V = 0.$$

3. Let $(V, \mathbf{A})$ denote the couple of potentials connected with the field excited in the vacuum by charges of densities ρ and $\mathbf{J}_\nu$. Then show that (*retarded potentials*)

$$V(x, y, z, t) = \frac{1}{4\pi\varepsilon_0}\iiint \frac{\rho(\xi, \eta, \zeta, t - r/c_0)}{r}\, d\xi d\eta d\zeta$$

$$\mathbf{A}(x, y, z, t) = \frac{\mu_0}{4\pi}\iiint \frac{\mathbf{J}(\xi, \eta, \zeta, t - r/c_0)}{r}\, d\xi d\eta d\zeta.$$

Here r stands the distance between the points (x, y, z) and (ξ, η, ζ) while the integrals are extended all over the space.

3.2 UNIVERSAL BOUNDARY CONDITIONS AND COMPATIBILITY RELATIONS

In what follows we will also consider the surfaces that model very thin slabs moving according to certain rule. So S may have a physical structure depending also on the time. Hence the coordinates of any point $P(x, y, z) \in S$ are functions of t, namely,

$$x = x(t), \qquad y = y(t), \qquad z = z(t). \tag{3.9}$$

If we differentiate the equation of S—that is, (3.5)—with respect to t, then we obtain

$$\frac{\partial}{\partial t}w + \mathbf{v} \cdot \operatorname{grad} w = 0, \tag{3.10a}$$

where $\mathbf{v}$ stands for the velocity of the point P:

$$\mathbf{v} = (dx/dt, dy/dt, dz/dt). \tag{3.10b}$$

By virtue of the fact that grad$w$ is perpendicular to S (i.e., parallel to $\mathbf{n}$) and shows the increasing direction of w, (3.10a) can also be written as

$$\frac{\partial}{\partial t}w = -(\mathbf{v} \cdot \mathbf{n})|\operatorname{grad} w|. \tag{3.10c}$$

In what follows we will replace $(\mathbf{v} \cdot \mathbf{n})$ and $|\operatorname{grad} w|$ by v_n and λ, respectively. Thus (3.10c) is also written as

$$\frac{\partial}{\partial t}w = -\lambda v_n, \qquad \lambda = |\operatorname{grad} w|, \qquad v_n = (\mathbf{v} \cdot \mathbf{n}) \tag{3.10d}$$

Reconsider now the function defined through (3.7) and compute its derivative with respect to t. By straightforward computations, one gets

$$\begin{aligned} \frac{\partial}{\partial t}\{f\} &= \frac{\partial f_2}{\partial t}H(w) + \frac{\partial f_1}{\partial t}[1 - H(w)] + f_2\frac{\partial H(w)}{\partial t} - f_1\frac{\partial H(w)}{\partial t} \\ &= \left\{\frac{\partial f}{\partial t}\right\} + [[f]]\frac{\partial H(w)}{\partial w}\frac{\partial w}{\partial t} \\ &= \left\{\frac{\partial f}{\partial t}\right\} - \lambda v_n[[f]]\delta(w). \end{aligned} \tag{3.11}$$

Let us insert (3.6a,b,...), (3.8b), and (3.11) into the equations (3.1a–d) and (3.3a,b). All the regular parts that are defined outside S are reduced to the expected relations that are satisfied in different regions separated by S (a trivial result!). As to the singular parts that are defined only on S: The coefficients of $\delta^{(k)}(w)(k = 0, 1, \ldots)$ taking place on both sides of an equation must be equal to each other. Thus, by using also (3.4a,b) for successive values of k, one gets the following set of equations.

$k = 0 \Rightarrow$

$$\lambda[[\mathbf{n} \times \mathbf{H}]] + \lambda v_n[[\mathbf{D}]] = -\mathrm{curl}\mathbf{H}_0 + \frac{\partial}{\partial t}\mathbf{D}_0 + \mathbf{J}_0, \tag{3.12a}$$

$$\lambda[[\mathbf{n} \times \mathbf{E}]] - \lambda v_n[[\mathbf{B}]] = -\mathrm{curl}\mathbf{E}_0 - \frac{\partial}{\partial t}\mathbf{B}_0, \tag{3.12b}$$

$$\lambda[[\mathbf{n} \cdot \mathbf{D}]] = \rho_0 - \mathrm{div}\mathbf{D}_0, \tag{3.12c}$$

$$\lambda[[\mathbf{n} \cdot \mathbf{B}]] = -\mathrm{div}\mathbf{B}_0, \tag{3.12d}$$

$$\lambda[[\mathbf{n} \cdot \mathbf{J}]] - \lambda v_n[[\rho]] = -\frac{\partial}{\partial t}\rho_0 - \mathrm{div}\mathbf{J}_0, \tag{3.12e}$$

$k \geq 1 \Rightarrow$

$$\mathrm{curl}\mathbf{H}_k + \lambda\mathbf{n} \times \mathbf{H}_{k-1} - \frac{\partial}{\partial t}\mathbf{D}_k + \lambda v_n \mathbf{D}_{k-1} = \mathbf{J}_k, \tag{3.13a}$$

$$\mathrm{curl}\mathbf{E}_k + \lambda\mathbf{n} \times \mathbf{E}_{k-1} + \frac{\partial}{\partial t}\mathbf{B}_k - \lambda v_n \mathbf{B}_{k-1} = 0, \tag{3.13b}$$

$$\mathrm{div}\mathbf{D}_k + \lambda\mathbf{n} \cdot \mathbf{D}_{k-1} = \rho_k, \tag{3.13c}$$

$$\mathrm{div}\mathbf{B}_k + \lambda\mathbf{n} \cdot \mathbf{B}_{k-1} = 0, \tag{3.13d}$$

$$\mathrm{div}\mathbf{J}_k + \lambda\mathbf{n} \cdot \mathbf{J}_{k-1} + \frac{\partial}{\partial t}\rho_k - \lambda v_n \rho_{k-1} = 0. \tag{3.13e}$$

The left-hand sides of the equations (3.12a–e) involve the limiting values of the field components on S, observed in different regions separated by S, while the right-hand sides involve only the singular parts of lowest order of the field components and source densities that are concentrated on S. The singular terms connected with the source densities (i.e., $\mathbf{J}_0$ and ρ_0) are all known beforehand through the definition of the source. As to the singular terms pertinent to the field (i.e., $\mathbf{E}_0, \mathbf{H}_0$, etc.), they can be determined through the relations (3.13a–e) in terms of the limiting values of the field components on S (by considering also the constitutive equations of S if it models a very thin material layer). This will be done later on in full detail (see Section 3.4.1 and Chapter 4). So the set of equations (3.12a–e), which are valid for all kinds of discontinuity surfaces and source distributions, are referred to as the *universal boundary conditions*. As to the set of equations (3.13a–e), which are written on S and interrelate the singular parts of different orders, they permit us to determine the singular parts of the field (i.e., $\mathbf{E}_k, \mathbf{H}_k$, etc.) starting from the known singular parts of the sources (i.e., $\mathbf{J}_k$ and ρ_k). Hence they are called the *compatibility conditions*. In what follows we will get further inside into the elaboration of them when we will investigate particular situations.

PROBLEMS

1. Show that (3.12c,d) reduce (3.12a,b) into

$$\lambda[[\mathbf{n} \times \mathbf{H}_t]] + \lambda v_n[[\mathbf{D}_t]] = -\mathrm{curl}\mathbf{H}_0 + \frac{\partial}{\partial t}\mathbf{D}_0 + \mathbf{J}_0 - v_n[\rho_0 - \mathrm{div}\mathbf{D_0}]\mathbf{n}$$

$$\lambda[[\mathbf{n} \times \mathbf{E}_t]] - \lambda v_n[[\mathbf{B}_t]] = -\mathrm{curl}\mathbf{E}_0 - \frac{\partial}{\partial t}\mathbf{B}_0 - v_n[\mathrm{div}\mathbf{B_0}]\mathbf{n}.$$

Here the sub-index t means the tangential component (i.e., projection on the tangent plane).

2. Discuss the dimensions of the terms that take place in (3.12a)–(3.13e).

3.2.1 An Example. Discontinuities on a Combined Sheet

In order to show how the compatibility conditions are elaborated in different particular cases, we want to consider a somewhat generalized version of the problem studied first in Panicali [17] and then in van Bladel [18] and Idemen [16]. The problem consists in finding the boundary conditions on a so-called combined sheet formed by a simple sheet and a double sheet lying (at rest) in the empty space. In this case we have

$$\rho = \rho_0\delta(w) + \rho_1\delta'(w), \qquad \mathbf{J} = \mathbf{J}_0\delta(w) + \mathbf{J}_1\delta'(w), \tag{3.14a,b}$$

and

$$\mathbf{P} = 0, \qquad \mathbf{M} = 0, \tag{3.14c}$$

which reduce the compatibility equations for $k \geq 2$ to

$$\mathrm{curl}\mathbf{H}_k + \lambda\mathbf{n} \times \mathbf{H}_{k-1} - \varepsilon_0\frac{\partial}{\partial t}\mathbf{E}_k = 0 \tag{3.15a}$$

$$\mathrm{curl}\mathbf{E}_k + \lambda\mathbf{n} \times \mathbf{E}_{k-1} + \mu_0\frac{\partial}{\partial t}\mathbf{H}_k = 0 \tag{3.15b}$$

$$\mathrm{div}\mathbf{E}_k + \lambda\mathbf{n} \cdot \mathbf{E}_{k-1} = 0 \tag{3.15c}$$

$$\mathrm{div}\mathbf{H}_k + \lambda\mathbf{n}\cdot\mathbf{H}_{k-1} = 0 \tag{3.15d}$$

$$\mathbf{n}\cdot\mathbf{J}_1 = 0 \tag{3.15e}$$

Since $\mathbf{E}_k \equiv 0$ and $\mathbf{H}_k \equiv 0$ for a certain finite index $k = N+1$, from (3.15a–d), one gets also

$$\mathbf{n}\times\mathbf{E}_N = 0, \qquad \mathbf{n}\cdot\mathbf{E}_N = 0, \qquad \mathbf{n}\times\mathbf{H}_N = \mathbf{0}, \qquad \mathbf{n}\cdot\mathbf{E}_N = 0.$$

Thus by induction one concludes

$$\mathbf{E}_k \equiv 0, \qquad \mathbf{H}_k \equiv 0, \qquad k = 1, 2, \ldots. \tag{3.16}$$

Now, by considering (3.16) in the remaining compatibility equations corresponding to $k = 1$, we write

$$\lambda\mathbf{n}\times\mathbf{H}_0 = \mathbf{J}_1, \tag{3.17a}$$

$$\lambda\mathbf{n}\times\mathbf{E}_0 = 0, \tag{3.17b}$$

$$\lambda\mathbf{n}\cdot\mathbf{E}_0 = \frac{\rho_1}{\varepsilon_0}, \tag{3.17c}$$

$$\lambda\mathbf{n}\cdot\mathbf{H}_0 = 0, \tag{3.17d}$$

$$\lambda\mathbf{n}\cdot\mathbf{J}_0 + \frac{\partial}{\partial t}\rho_1 + \mathrm{div}\mathbf{J}_1 = 0, \tag{3.17e}$$

which yield

$$\mathbf{E}_0 = \frac{1}{\lambda}\frac{\rho_1}{\varepsilon_0}\mathbf{n}, \qquad \mathbf{H}_0 = -\frac{1}{\lambda}\mathbf{n}\times\mathbf{J}_1. \tag{3.18a,b}$$

By substituting these expressions of $\mathbf{E}_0$ and $\mathbf{H}_0$ in the universal relations (3.12a–e), we get

$$\lambda[[\mathbf{n}\times\mathbf{H}]] = \mathrm{curl}\left\{\frac{1}{\lambda}\mathbf{n}\times\mathbf{J}_1\right\} + \frac{\varepsilon_0}{\lambda}\frac{\partial}{\partial t}\left\{\frac{\rho_1}{\varepsilon_0}\mathbf{n}\right\} + \mathbf{J}_0 \tag{3.19a}$$

$$\lambda[[\mathbf{n}\times\mathbf{E}]] = -\mathrm{curl}\left\{\frac{1}{\lambda}\frac{\rho_1}{\varepsilon_0}\mathbf{n}\right\} + \frac{\mu_0}{\lambda}\frac{\partial}{\partial t}\{\mathbf{n}\times\mathbf{J}_1\} \tag{3.19b}$$

$$\lambda[[\mathbf{n}\cdot\mathbf{D}]] = \rho_0 - \varepsilon_0\mathrm{div}\left\{\frac{1}{\lambda}\frac{\rho_1}{\varepsilon_0}\mathbf{n}\right\} \tag{3.19c}$$

$$\lambda[[\mathbf{n}\cdot\mathbf{B}]] = \mu_0\mathrm{div}\left\{\frac{1}{\lambda}\mathbf{n}\times\mathbf{J}_1\right\} \tag{3.19d}$$

$$\lambda[[\mathbf{n}\cdot\mathbf{J}]] = -\frac{\partial}{\partial t}\rho_0 - \mathrm{div}\mathbf{J}_0. \tag{3.19e}$$

When S consists of a plane, then one has $\lambda = 1$, which reduces the above equations into those derived in references 16–18. Note that in the case when S is a sphere or circular cylinder, one has also $\lambda = 1$. That means that the formulas derived in references 16–18 are valid for these kinds of boundaries also.

3.3 THE CONCEPT OF MATERIAL SHEET

Consider a very thin layer of width $(2d)$ about a regular surface S (see Fig. 3.3). The layer is supposed to be formed by a *linear* material whose constitutive parameters are constant (i.e., it is homogeneous, local, and time invariant).It may be isotropic or anisotropic. If it is anisotropic, its properties are the same in tangential directions. In other words, it seems isotropic when one observes it in directions parallel to the tangent planes of S. So, its constitutive parameters in tangential and normal directions differ from each other. In what follows they will be denoted by $(\varepsilon, \mu, \sigma)$ and $(\varepsilon_n, \mu_n, \sigma_n)$ in tangential and normal directions, respectively.* The regions above and below the layer are supposed to be simple with constitutive parameters $(\varepsilon_2, \mu_2, \sigma_2)$ and $(\varepsilon_1, \mu_1, \sigma_1)$, respectively. If the width of the layer is sufficiently small as compared with both the wavelength of the electromagnetic fields we will consider and the curvature of the slab, then in our discussions we can replace it by S. But to introduce the effect of the charges, currents, and polarizations (both electric and magnetic) excited inside the slab, we have to assign to S certain constitutive parameters. These latter are determined by considering the averages of the field components inside the slab (see Section 4.3.1 below). A surface S having this kind

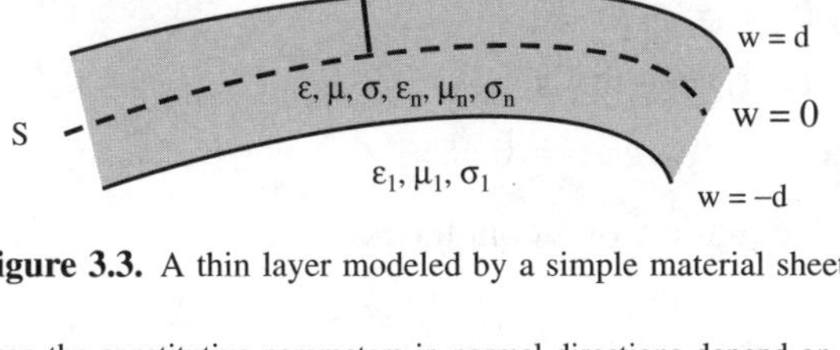

Figure 3.3. A thin layer modeled by a simple material sheet.

*For the case when the constitutive parameters in normal directions depend on the depth, see Section 4.3.1.B.

of constitutive parameters is referred to as a *simple material sheet*. In boundary-value problems, material sheets are represented by the jumps of the field components to be observed on them. These jumps constitute the boundary conditions on the material sheet in question. Now we want to derive the boundary conditions on a simple material sheet.

Because of the charges, currents, and polarizations to be excited inside the layer modeled by S, the densities $\rho_0, \mathbf{J}_0, \mathbf{P}_0$, and $\mathbf{M}_0$, which are all concentrated on S, may be different from zero. Furthermore, owing to the possible differences between the charges accumulated on the surfaces $w = d$ and $w = (-d)$, the slab may seem as a double sheet. That means that the dipole density ρ_1 may also be different from zero. Because of the motions of the charges and S, in addition to ρ_1, the density corresponding to dipole currents (i.e., $\mathbf{J}_1$) may also differ from zero. All the remaining terms existing in the compatibility relations (3.13a–e) are naught, which permits us to write them as follows:

$k = 1 \Rightarrow$

$$\lambda \mathbf{n} \times \mathbf{H}_0 + \lambda v_n \mathbf{D}_0 = \varepsilon_0 \frac{\partial}{\partial t} \mathbf{E}_1 - \text{curl}\mathbf{H}_1 + \mathbf{J}_1 \tag{3.20a}$$

$$\lambda \mathbf{n} \times \mathbf{E}_0 - \lambda v_n \mathbf{B}_0 = -\mu_0 \frac{\partial}{\partial t} \mathbf{H}_1 - \text{curl}\mathbf{E}_1 \tag{3.20b}$$

$$\lambda \mathbf{n} \cdot \mathbf{D}_0 = \rho_1 - \varepsilon_0 \text{div}\mathbf{E}_1 \tag{3.20c}$$

$$\lambda \mathbf{n} \cdot \mathbf{B}_0 = -\mu_0 \text{div}\mathbf{H}_1 \tag{3.20d}$$

$$\lambda \mathbf{n} \cdot \mathbf{J}_0 - \lambda v_n \rho_0 = -\frac{\partial}{\partial t} \rho_1 - \text{div}\mathbf{J}_1 \tag{3.20e}$$

$k \geq 2 \Rightarrow$

$$\lambda \mathbf{n} \times \mathbf{H}_{k-1} + \lambda \varepsilon_0 v_n \mathbf{E}_{k-1} = \varepsilon_0 \frac{\partial}{\partial t} \mathbf{E}_k - \text{curl}\mathbf{H}_k \tag{3.21a}$$

$$\lambda \mathbf{n} \times \mathbf{E}_{k-1} - \lambda \mu_0 v_n \mathbf{H}_{k-1} = -\mu_0 \frac{\partial}{\partial t} \mathbf{H}_k - \text{curl}\mathbf{E}_k \tag{3.21b}$$

$$\lambda \mathbf{n} \cdot \mathbf{E}_{k-1} = -\text{div}\mathbf{E}_k \tag{3.21c}$$

$$\lambda \mathbf{n} \cdot \mathbf{H}_{k-1} = -\text{div}\mathbf{H}_k, \tag{3.21d}$$

$$\mathbf{n} \cdot \mathbf{J}_{k-1} - v_n \rho_{k-1} = 0. \tag{3.21e}$$

From the last equation one concludes

$$J_{kn} = \rho_k v_n, \qquad k = 1, 2, \ldots, \tag{3.22}$$

which accords with the general definition of current.

By repeating the reasoning made in Section 3.2.1, one gets from (3.21a–d) that

$$\mathbf{E}_k = 0, \qquad \mathbf{H}_k = 0, \qquad k = 1, 2, \ldots . \tag{3.23}$$

Indeed, for a certain finite index $k = N+1$ one has $\mathbf{E}_{N+1} \equiv 0$ and $\mathbf{H}_{N+1} \equiv 0$. Hence for $k = N+1$ (3.21a–d) are reduced to

$$\mathbf{n} \times \mathbf{H}_N + \varepsilon_0 v_n \mathbf{E}_N = 0,$$
$$\mathbf{n} \times \mathbf{E}_N - \mu_0 v_n \mathbf{H}_N = 0$$

and

$$\mathbf{n} \cdot \mathbf{E}_N = 0, \qquad \mathbf{n} \cdot \mathbf{H}_N = 0.$$

If we consider the cross-multiplication of the first equation with $\mathbf{n}$ and make a simplification by using the second and fourth, then we obtain

$$[1 - (v_n/c_0)^2]\mathbf{H}_N = 0, \tag{3.24}$$

where $c_0 = 1/\sqrt{\varepsilon_0 \mu_0}$ stands, as usual, for the velocity of the electromagnetic wave in vacuum. Since always one has $v_n < c_0$, (3.24) gives $\mathbf{H}_N = 0$.

Similarly, one can show also that $\mathbf{E}_N = 0$. Thus by induction one obtains (3.23). Now let us return to (3.20a–e) and make possible simplifications by considering (3.23), to get

$$\lambda \mathbf{n} \times \mathbf{H}_0 + \lambda v_n \mathbf{D}_0 = \mathbf{J}_1, \tag{3.25a}$$
$$\lambda \mathbf{n} \times \mathbf{E}_0 - \lambda v_n \mathbf{B}_0 = 0, \tag{3.25b}$$
$$\lambda \mathbf{n} \cdot \mathbf{D_0} = \rho_1, \tag{3.25c}$$
$$\lambda \mathbf{n} \cdot \mathbf{B}_0 = 0, \tag{3.25d}$$
$$\lambda \mathbf{n} \cdot \mathbf{J}_0 - \lambda v_n \rho_0 = -\frac{\partial}{\partial t}\rho_1 - \operatorname{div}\mathbf{J}_1. \tag{3.25e}$$

These equations consist of the compatibility conditions to be considered together with the boundary conditions (3.12a–e) on a simple material sheet. Their dependence on (λv_n) shows that the structure of the function $w(x, y, z, t)$ effects deeply the analysis of the boundary conditions satisfied by $[[\mathbf{n} \times \mathbf{E}]]$, $[[\mathbf{n} \times \mathbf{H}]]$, $[[\mathbf{D}]]$, $[[\mathbf{n} \cdot \mathbf{D}]]$, $[[\mathbf{B}]]$, and $[[\mathbf{n} \cdot \mathbf{B}]]$. So, in what follows we will consider in full detail some particular cases that have important practical applications. The particular cases in question are as follows:

i. The sheet S is at rest; that is, $w = w(x, y, z)$ (see Chapter 4).

ii. The sheet is in motion; that is, $w = w(x, y, z, t)$ (see Chapter 5).

PROBLEMS

1. Show that when $\mathbf{E}_1 = 0, \mathbf{H}_1 = 0, \mathbf{P}_1 = 0$, and $\mathbf{M}_1 = 0$, the normal and tangential components of (3.13a,b) give, respectively, $J_{1n} = \rho_1 v_n$ and

$$\lambda \mathbf{n} \times \mathbf{H}_{0t} + \lambda v_n \mathbf{D}_{0t} = \mathbf{J}_{1t}, \qquad \lambda \mathbf{n} \times \mathbf{E}_{0t} - \lambda v_n \mathbf{B}_{0t} = 0.$$

From these last two equations, one gets also

$$\lambda^2 \left[1 - \frac{(v \cdot n)^2}{c_0^2} \right] \mathbf{E}_{0t} = -\lambda \mu_0 v_n \mathbf{J}_{1t} + \mu_0 (\lambda v_n)^2 \mathbf{P}_{0t} - \lambda^2 v_n \mathbf{n} \times \mathbf{M}_{0t}$$

$$\lambda^2 \left[1 - \frac{(v \cdot n)^2}{c_0^2} \right] \mathbf{H}_{0t} = \varepsilon_0 (\lambda v_n)^2 \mathbf{M}_{0t} - \lambda \mathbf{n} \times \mathbf{J}_{1t} + \lambda^2 v_n \mathbf{n} \times \mathbf{P}_{0t}.$$

Here the sub-indices n and t refer to the projections on the normal vector and tangent plane, respectively, and $1/c_0^2 = \varepsilon_0 \mu_0$.

3.4 THE CASE OF MONOCHROMATIC FIELDS

The fields excited in practical applications are very frequently time-harmonic (i.e., *monochromatic*). In such a case, all components of the field are changed with time as a simple trigonometric function with a certain frequency. One writes, for example, for the charge density

$$\rho(x, y, z, t) = \rho^*(x, y, z) \cos\{\omega t - \alpha(x, y, z)\}. \qquad (3.26a)$$

Here $\rho^*(x, y, z)$, which is supposed to be *positive*, is the amplitude of the charge density while the *positive* constant ω (or $\omega/2\pi = f$) and the term $\alpha(x, y, z)$ are referred to as the *frequency* and *phase* of the charge density, respectively. The main advantage provided by the monochromatic field assumption in theoretical investigations is that the time dependence is completely known (and fixed) beforehand, which enormously facilitates investigations. It is worthwhile to remark here that this assumption will not, in general, cause to severe restrictions for the results to be obtained because any field can, in general, be represented by a Fourier series or integral in terms of trigonometric functions.

An expression of type (3.26a) can also be written as

$$\rho(x, y, z, t) = \Re\{\rho^*(x, y, z) e^{i\alpha(x,y,z,)} e^{-i\omega t}\}, \qquad (3.26b)$$

where $\Re$ stands for the *real part*. By combining the exponential phase factor $e^{i\alpha(x,y,z)}$ with the real-valued amplitude $\rho^*(x, y, z)$, one writes (3.26b) as follows:

$$\rho(x,y,z,t) = \Re\{\rho(x,y,z)e^{-i\omega t}\}. \tag{3.26c}$$

Here $\rho(x,y,z)$ stands for the product $[\rho^*(x,y,z)e^{i\alpha(x,y,z)}]$ and has, in general, a complex value. $\rho(x,y,z)$ is referred to as the *complex amplitude* of the charge density. It is worthwhile to draw the attention to the fact that we use the same letter ρ to denote both $\rho(x,y,z,t)$ and $\rho(x,y,z)$ because there will be no confusion since the first is real-valued and depends also on t while the second is of complex-valued and does not depend on t. If the terms taking place in the Maxwell equations as well as those in (3.2) and (3.4a,b) are all replaced by expressions similar to (4.10b), then one gets

$$\Re\left\{\left[\text{curl}\mathbf{H} + i\omega\left(\varepsilon + \frac{i\sigma}{\omega}\right)\mathbf{E} - \mathbf{J}_\nu\right]e^{-i\omega t}\right\} = 0,$$
$$\Re\left\{[\text{curl}\mathbf{E} - i\omega\mu\mathbf{H}]\,e^{-i\omega t}\right\} = 0$$

and

$$\Re\{[\text{div}(\varepsilon\mathbf{E}) - \rho]e^{-i\omega t}\} = 0, \qquad \Re\{\text{div}(\mu\mathbf{H})e^{-i\omega t}\} = 0.$$

Since these equations are to be satisfied for all $t \in (-\infty, \infty)$, one writes also

$$\text{curl}\mathbf{H} + i\omega\varepsilon'\mathbf{E} = \mathbf{J}_\nu, \qquad \text{curl}\mathbf{E} - i\omega\mu\mathbf{H} = 0, \tag{3.27a,b}$$
$$\text{div}(\varepsilon\mathbf{E}) = \rho, \qquad \text{div}(\mu\mathbf{H}) = 0, \tag{3.27c,d}$$

where one has

$$\varepsilon' = \left(\varepsilon + \frac{i\sigma}{\omega}\right). \tag{3.28}$$

The complex-valued ε' is referred to as the complex permittivity of the medium. The equations in (3.27a–d), which are satisfied by the complex amplitudes of a monochromatic field, are referred to as the complex Maxwell equations. By taking the divergences of (3.27a,b), one gets

$$i\omega\text{div}(\varepsilon'\mathbf{E}) = \text{div}\mathbf{J}_\nu, \qquad \text{div}(\mu\mathbf{H}) = 0. \tag{3.29}$$

When v_n is equal to naught (see Chapter 4) or does *not change with time* on the boundary (see Chapter 5), for a monochromatic field the universal boundary conditions (3.12a–e) and compatibility relations (3.13a–e) are reduced to the following:

$k = 0 \Rightarrow$

$$\lambda[[\mathbf{n} \times \mathbf{H}]] + \lambda v_n[[\mathbf{D}]] = -\mathrm{curl}\mathbf{H}_0 - i\omega\mathbf{D}_0 + \mathbf{J}_0, \tag{3.30a}$$

$$\lambda[[\mathbf{n} \times \mathbf{E}]] - \lambda v_n[[\mathbf{B}]] = -\mathrm{curl}\mathbf{E}_0 + \mathrm{i}\omega\mathbf{B}_0, \tag{3.30b}$$

$$\lambda[[\mathbf{n} \cdot \mathbf{D}]] = \rho_0 - \mathrm{div}\mathbf{D}_0, \tag{3.30c}$$

$$\lambda[[\mathbf{n} \cdot \mathbf{B}]] = -\mathrm{div}\mathbf{B}_0, \tag{3.30d}$$

$$\lambda[[\mathbf{n} \cdot \mathbf{J}]] - \lambda v_n[[\rho]] = i\omega\rho_0 - \mathrm{div}\mathbf{J}_0, \tag{3.30e}$$

$k \geq 1 \Rightarrow$

$$\mathrm{curl}\mathbf{H}_k + \lambda\mathbf{n} \times \mathbf{H}_{k-1} + i\omega\mathbf{D}_k + \lambda v_n\mathbf{D}_{k-1} = \mathbf{J}_k, \tag{3.31a}$$

$$\mathrm{curl}\mathbf{E}_k + \lambda\mathbf{n} \times \mathbf{E}_{k-1} - i\omega\mathbf{B}_\mathrm{k} - \lambda v_n\mathbf{B}_{k-1} = 0, \tag{3.31b}$$

$$\mathrm{div}\mathbf{D}_k + \lambda\mathbf{n} \cdot \mathbf{D}_{k-1} = \rho_k, \tag{3.31c}$$

$$\mathrm{div}\mathbf{B}_k + \lambda\mathbf{n} \cdot \mathbf{B}_{k-1} = 0, \tag{3.31d}$$

$$\mathrm{div}\mathbf{J}_k + \lambda\mathbf{n} \cdot \mathbf{J}_{k-1} - i\omega\rho_k - \lambda v_n\rho_{k-1} = 0. \tag{3.31e}$$

3.4.1 Discontinuities on the Interface Between Two Simple Media that Are at Rest

Let S denote the interface between two simple media which are at rest (see Fig. 3.4). Since S has no physical existence, one has

$$v_n \equiv 0, \qquad \mathbf{P}_k \equiv 0, \qquad \mathbf{M}_k \equiv 0, \tag{3.32a}$$

$$\mathbf{D}_k \equiv \varepsilon_0\mathbf{E}_k, \qquad \mathbf{B}_k \equiv \mu_0\mathbf{H}_k, \qquad k = 1, 2, \ldots. \tag{3.32b}$$

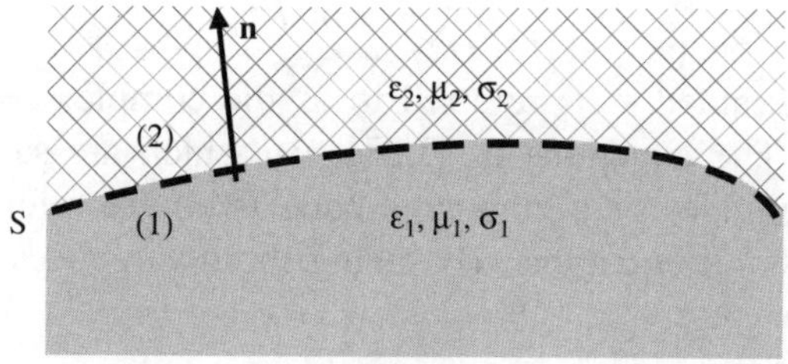

Figure 3.4. Interface between two simple media.

Furthermore, except ρ_0 and $\mathbf{J}_0$, all charge and current coefficients are also equal to naught. ρ_0, when it differs from zero, denotes the density of charge accumulated on S because of the conductances of the regions separated by S. Thus the compatibility relations (3.31a–e) are reduced to the following:

$$\mathrm{curl}\mathbf{H}_k + \lambda\mathbf{n} \times \mathbf{H}_{k-1} + i\omega\varepsilon_0\mathbf{E}_k = 0, \tag{3.33a}$$

$$\mathrm{curl}\mathbf{E}_k + \lambda\mathbf{n} \times \mathbf{E}_{k-1} - i\omega\mu_0\mathbf{H}_k = 0, \tag{3.33b}$$

$$\mathrm{div}\mathbf{E}_k + \lambda\mathbf{n} \cdot \mathbf{E}_{k-1} = 0, \tag{3.33c}$$

$$\mathrm{div}\mathbf{H}_k + \lambda\mathbf{n} \cdot \mathbf{H}_{k-1} = 0, \tag{3.33d}$$

$$\mathrm{div}\mathbf{J}_k + \lambda\mathbf{n} \cdot \mathbf{J}_{k-1} = 0. \tag{3.33e}$$

From these equations one gets

$$\mathbf{E}_k \equiv 0, \qquad \mathbf{H}_k \equiv 0, \qquad k = 0, 1, \ldots. \tag{3.34}$$

If one uses these results in the boundary relations (3.30a–e), then one gets

$$\lambda[[\mathbf{n} \times \mathbf{H}]] = \mathbf{J}_0, \tag{3.35a}$$

$$\lambda[[\mathbf{n} \times \mathbf{E}]] = 0, \tag{3.35b}$$

$$\lambda[[\mathbf{n} \cdot \mathbf{D}]] = \rho_0, \tag{3.35c}$$

$$\lambda[[\mathbf{n} \cdot \mathbf{B}]] = 0, \tag{3.35d}$$

$$\lambda[[\mathbf{n} \cdot \mathbf{J}]] = i\omega\rho_0 - \mathrm{div}\mathbf{J}_0. \tag{3.35e}$$

From *experiments* it is known that when one of the regions separated by S is not *perfectly conducting*, one always has $\mathbf{J}_0 \equiv 0$. By considering this in (3.35e), we obtain

$$\rho_0 = -i\left(\frac{\lambda}{\omega}\right)[[\mathbf{n} \cdot \mathbf{J}]], \qquad \sigma_{\mathrm{j}} \neq \infty, \ j = 1, 2 \tag{3.36a}$$

If one inserts $\mathbf{J}_0 \equiv 0$ and (3.36a) into (3.35a–d), then one gets finally

$$[[\mathbf{n} \times \mathbf{H}]] = 0, \qquad [[\mathbf{n} \times \mathbf{E}]] = 0 \tag{3.36b,c}$$

and

$$[[\mathbf{n} \cdot \mathbf{D} + \frac{i}{\omega}\mathbf{n} \cdot \mathbf{J}]] = 0, \qquad [[\mathbf{n} \cdot \mathbf{B}]] = 0. \tag{3.36d,e}$$

The last two equation are also written, more explicitly, as follows:

$$\varepsilon_1' E_n^{(1)} = \varepsilon_2' E_n^{(2)}, \qquad \mu_1 H_n^{(1)} = \mu_2 H_n^{(2)}, \tag{3.37a,b}$$

where

$$\varepsilon_j' = \varepsilon_j + i\sigma_j/\omega, \qquad j = 1, 2. \tag{3.37c}$$

ε_j' defined in (3.37c) is referred to as the complex permittivity of the region j.

PROBLEMS

1. Show that (3.37a,b) are equivalent to the singular parts of the following identities considered in the complex form:

$$0 = \operatorname{div}\operatorname{curl}\mathbf{H} = -i\omega\operatorname{div}(\varepsilon'\mathbf{E}), \qquad 0 = \operatorname{div}\operatorname{curl}\mathbf{E} = i\omega\operatorname{div}(\mu\mathbf{H}).$$

2. Show that (3.37a,b) can also be written as

$$\frac{\partial}{\partial t}D_n^{(1)} + J_n^{(1)} = \frac{\partial}{\partial t}D_n^{(2)} + J_n^{(2)}, \qquad \frac{\partial}{\partial t}B_n^{(1)} = \frac{\partial}{\partial t}B_n^{(2)}$$

which consist of the singular parts of

$$\operatorname{div}\operatorname{curl}\mathbf{H} = 0, \qquad \operatorname{div}\operatorname{curl}\mathbf{E} = 0$$

considered in the real form.

CHAPTER 4

Boundary Conditions on Material Sheets at Rest

4.1 UNIVERSAL BOUNDARY CONDITIONS AND COMPATIBILITY RELATIONS FOR A FIXED MATERIAL SHEET

When the surface S considered in previous sections is at rest, the function $w(x, y, z, t)$ becomes independent of the time parameter t and yields $\partial w/\partial t \equiv 0$. Furthermore, in this case one has also $\mathbf{J}_1 = 0$ because there will be no current of dipoles inside the simple slab modeled by S. By canceling these terms from the boundary and compatibility relations derived in previous sections, one gets the following rather simple equations (see (3.12a–e) and (3.25a–e)):

$$\lambda[[\mathbf{n} \times \mathbf{H}]] = -\mathrm{curl}\mathbf{H}_0 + \frac{\partial}{\partial t}\mathbf{D}_0 + \mathbf{J}_0 \tag{4.1a}$$

$$\lambda[[\mathbf{n} \times \mathbf{E}]] = -\mathrm{curl}\mathbf{E}_0 - \frac{\partial}{\partial t}\mathbf{B}_0 \tag{4.1b}$$

$$\lambda[[\mathbf{n} \cdot \mathbf{D}]] = \rho_0 - \mathrm{div}\mathbf{D}_0 \tag{4.1c}$$

$$\lambda[[\mathbf{n} \cdot \mathbf{B}]] = -\mathrm{div}\mathbf{B}_0 \tag{4.1d}$$

$$\lambda[[\mathbf{n} \cdot \mathbf{J}]] = -\frac{\partial}{\partial t}\rho_0 - \mathrm{div}\mathbf{J}_0 \tag{4.1e}$$

Discontinuties in the Electromagnetic Field, First Edition. By M. Mithat Idemen.

Published 2011 by John Wiley & Sons, Inc.

and

$$\mathbf{n} \times \mathbf{H}_0 = 0 \tag{4.2a}$$

$$\mathbf{n} \times \mathbf{E}_0 = 0 \tag{4.2b}$$

$$\lambda \mathbf{n} \cdot \mathbf{D_0} = \rho_1 \tag{4.2c}$$

$$\lambda \mathbf{n} \cdot \mathbf{B}_0 = 0 \tag{4.2d}$$

$$\lambda \mathbf{n} \cdot \mathbf{J}_0 = -\frac{\partial}{\partial t} \rho_1 \tag{4.2e}$$

The reason for $\rho_1 \neq 0$ is the possibility of difference between the charge densities accumulated on different sides of S, which makes the material sheet a combination of simple and double layers. It is seen from (4.2e) that the double layer to be formed on the sheet is a consequence of the conduction in normal direction inside the slab. If the slab is not conducting in normal direction (i.e., when $\sigma_n = 0$), then $\rho_1 = 0$, which claims that S consists of a simple sheet only.

In accordance with (4.2a–d), one has

$$\begin{aligned}
\text{curl}\mathbf{E}_0 &= \text{curl}(E_{0n}\mathbf{n}) \\
&= \text{curl}\left(\frac{1}{\lambda} E_{0n} \text{grad} w\right) \\
&= -\text{grad} w \times \text{grad}\left(\frac{1}{\lambda} E_{0n}\right) \\
&= -\lambda \mathbf{n} \times \text{grad}\left(\frac{1}{\lambda} E_{0n}\right)
\end{aligned} \tag{4.3}$$

and similarly

$$\text{curl}\mathbf{H}_0 = -\lambda \mathbf{n} \times \text{grad}\left(\frac{1}{\lambda} H_{0n}\right), \tag{4.4}$$

which reduce (4.1a–d) to

$$\frac{\partial}{\partial t}\{\lambda[[\mathbf{n} \times \mathbf{H}]]\} = \frac{\partial}{\partial t}\left\{\mathbf{J}_{0t} + \frac{\partial}{\partial t}\mathbf{P}_{\text{ot}}\right\} - \frac{\lambda}{\mu_0}\mathbf{n} \times \text{grad}\left\{\frac{1}{\lambda}\left(\frac{\partial}{\partial t} M_{0n}\right)\right\} \tag{4.5a}$$

$$\frac{\partial}{\partial t}\{\lambda[[\mathbf{n} \times \mathbf{E}]]\} = -\frac{\partial}{\partial t}\left\{\frac{\partial}{\partial t}\mathbf{M}_{\text{ot}}\right\} - \frac{\lambda}{\varepsilon_0}\mathbf{n} \times \text{grad}\left\{\frac{1}{\lambda}\left(J_{0n} + \frac{\partial}{\partial t} P_{0n}\right)\right\} \tag{4.5b}$$

$$\lambda[[\mathbf{n} \cdot \mathbf{D}]] = \rho_0 - \mathrm{div}\left(\frac{\rho_1}{\lambda}\mathbf{n}\right) - \mathrm{div}\mathbf{P}_{0t} \tag{4.5c}$$

$$\lambda[[\mathbf{n} \cdot \mathbf{B}]] = -\mathrm{div}\mathbf{M}_{0t}. \tag{4.5d}$$

Note that in deriving these relations we have taken into consideration the fact that the normal component of $(\mathbf{J}_0 + \partial\mathbf{D}_0/\partial t)$ as well as the tangential components of $\mathbf{E}_0$ and $\mathbf{H}_0$ are always equal to naught because of (4.2 a–e) and $\partial\lambda/\partial t \equiv 0$. On the other hand, E_{0n} and H_{0n} appearing in (4.3) and (4.4) were replaced by their expressions obtained through (4.2c,d).

Equations (4.5a–d) constitute the most general form of the boundary relations to be satisfied on a simple motionless sheet capable of supporting surface charges and dipoles. The jump discontinuities that take place on the left-hand sides are given by the differences of the field values observed on both sides of S, say $\mathbf{E}^{1,2}$ and $\mathbf{H}^{1,2}$. One has, for example, $[[\mathbf{H}]] = \mathbf{H}^2 - \mathbf{H}^1$. The terms lying on the right-hand sides of these equations are connected to the limiting field values $\mathbf{E}^{1,2}$ and $\mathbf{H}^{1,2}$, observed on different sides of S, through the constitutive equations of the sheet. These latter are derived in Section 4.3 below.

PROBLEM

1. Show that one has

$$\mathrm{div}\mathbf{E}_0 = \mathrm{div}(E_{\mathrm{on}}\mathbf{n}) = \mathrm{div}\left[\frac{1}{\varepsilon_0}\left(\frac{\rho_1}{\lambda} - P_{\mathrm{on}}\right)\mathbf{n}\right],$$

$$\mathrm{div}\mathbf{H}_0 = \mathrm{div}(H_{\mathrm{on}}\mathbf{n}) = -\mathrm{div}\left[\left(\frac{1}{\mu_0}M_{\mathrm{on}}\right)\mathbf{n}\right].$$

4.2 SOME GENERAL RESULTS [19]

Before going further to the boundary relations connected with particular cases, we want to draw attention to some general points to be derived from the expressions of the general relations (4.5a–d).

1. From (4.5a) one concludes that the source of discontinuity of the tangential magnetic field is not merely the tangential electric current $\mathbf{J}_{ot}$, as accepted by almost everybody. The existence of the tangential dielectric polarization $\mathbf{P}_{0t}$ as well as the nonuniformity of the normal magnetic polarization M_{0n} divided by λ contribute

also to this discontinuity. Note that λ is connected closely to the geometry (curvature) of the sheet S.

2. Similarly, (4.5b) shows that among the causes of the discontinuity of the tangential electric field there are the tangential magnetic polarization $\mathbf{M}_{0t}$, the normal electric current J_{0n}, and the normal polarization P_{0n} as well as the geometrical parameter λ connected with the sheet S. This point is very important especially when one tries to interpret the discontinuities on impedance-type boundaries (see Section 4.3.1.E below).
3. The result to be derived from (4.5d) is much more interesting: The normal component of the magnetic induction can also be discontinuous if the divergence of the tangential magnetic polarization on S is different from zero.

4.3 SOME PARTICULAR CASES

Now we want to consider the following particular cases in some detail:

A. The case of planar S

B. The case of circular cylindrical or spherical S

C. The case of rotationally symmetric conical S

4.3.1 Planar Material Sheet Between Two Simple Media

When S consists of a plane, say the plane $z = 0$, the function $w(x, y, z)$, the unit vector $\mathbf{n}$ and λ, mentioned in Section 3.3 above, are reduced to

$$w = z, \qquad \mathbf{n} = \mathbf{e}_z, \qquad \lambda = 1. \tag{4.6}$$

In this case the boundary relations (4.5a–d) are read as follows (see Idemen [19, p. 671, formula (4.3a–d)]:

$$\frac{\partial}{\partial t}[[\mathbf{n} \times \mathbf{H}]] = \frac{\partial}{\partial t}\left\{\mathbf{J}_{0t} + \frac{\partial}{\partial t}\mathbf{P}_{0t}\right\} - \frac{1}{\mu_0}\mathbf{n} \times \text{grad}\left\{\frac{\partial}{\partial t}M_{0n}\right\} \tag{4.7a}$$

$$\frac{\partial}{\partial t}[[\mathbf{n} \times \mathbf{E}]] = -\frac{\partial}{\partial t}\left\{\frac{\partial}{\partial t}\mathbf{M}_{0t}\right\} - \frac{1}{\varepsilon_0}\mathbf{n} \times \text{grad}\left\{\mathrm{J}_{0n} + \frac{\partial}{\partial t}\mathrm{P}_{0n}\right\} \tag{4.7b}$$

$$[[\mathbf{n} \cdot \mathbf{D}]] = \rho_0 - \operatorname{div}(\rho_1 \mathbf{n}) - \operatorname{div}\mathbf{P}_{0t} \tag{4.7c}$$
$$[[\mathbf{n} \cdot \mathbf{B}]] = -\operatorname{div}\mathbf{M}_{0t}. \tag{4.7d}$$

Now suppose that S models a very thin slab of width $(2d)$ filled with an anisotropic material mentioned in Section 3.3 above (see Fig. 4.1). If one denotes the limiting values of any field component, say $\mathbf{E}$, for $z \to (d+0)$ and $z \to (-d-0)$ by $\mathbf{E}^{(2)}$ and $\mathbf{E}^{(1)}$, respectively, then the jump discontinuity mentioned in (4.7a–d) becomes

$$[[\mathbf{n} \times \mathbf{E}]] = \mathbf{e}_z \times [\mathbf{E}^{(2)} - \mathbf{E}^{(1)}] \tag{4.8a}$$

with

$$\mathbf{E}^{(1)} = \mathbf{E}(x, y, -d-0), \qquad \mathbf{E}^{(2)} = \mathbf{E}(x, y, d+0). \tag{4.8b}$$

It is quite plausible to assume that inside S the tangential components $(\mathbf{J}_{0t}, \mathbf{P}_{0t})$ are proportional to $(\mathbf{E}_t^{(1)}, \mathbf{E}_t^{(2)})$ while the normal components (J_{0n}, P_{0n}) are proportional to $(E_n^{(1)}, E_n^{(2)})$. Hence we can suppose that the following relations are valid with known coefficients ($\equiv$ constitutive parameters of S):

$$\mathbf{J}_{0t} = \sigma^*{}_1 \mathbf{E}_t^{(1)} + \sigma^*{}_2 \mathbf{E}_t^{(2)}, \tag{4.9a}$$
$$\mathbf{P}_{0t} = \varepsilon_0 \chi_1^e \mathbf{E}_t^{(1)} + \varepsilon_0 \chi_2^e \mathbf{E}_t^{(2)}, \tag{4.9b}$$
$$\mathbf{M}_{0t} = \mu_0 \chi_1^m \mathbf{H}_t^{(1)} + \mu_0 \chi_2^m \mathbf{H}_t^{(2)}, \tag{4.9c}$$
$$J_{0n} = \sigma_{1n} E_n^{(1)} + \sigma_{2n} E_n^{(2)}, \tag{4.9d}$$
$$P_{0n} = \varepsilon_0 \chi_{1n}^e E_n^{(1)} + \varepsilon_0 \chi_{2n}^e E_n^{(2)}, \tag{4.9e}$$
$$M_{0n} = \mu_0 \chi_{1n}^m H_n^{(1)} + \mu_0 \chi_{2n}^m H_n^{(2)}. \tag{4.9f}$$

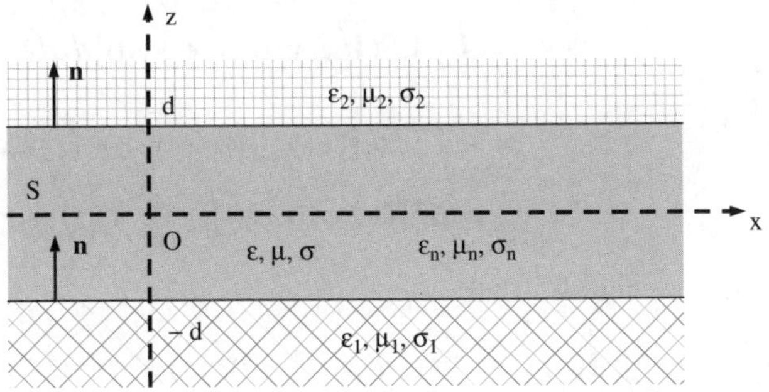

Figure 4.1. A planar material sheet simulating an anisotropic thin layer.

To derive explicit expressions of the constitutive parameters for a sheet simulating a given thin planar slab, let us consider the case of *monochromatic* fields of frequency ω (see Section 3.4) and assume that the thickness $(2d)$ of the layer is very small as compared to the free-space wavelength of the field, that is, $(2d) \ll 2\pi/(\omega\sqrt{\varepsilon_0\mu_0})$. In this case the distribution generated by the current density $\mathbf{J}(x, y, z)$ existing inside the layer $(-d) < z < d$, for example, is written as

$$< \mathbf{J}(x, y, z), \varphi(x, y, z) > = \iiint \mathbf{J}(x, y, z)\varphi(x, y, z)\, dxdydz, \quad (4.10a)$$

where $\varphi(x, y, z)$ stands for any test function. The last expression is based on the fact that

$$\mathbf{J}(x, y, z) \in L_0^2(R^3) \subset L_0^1(R^3) \quad (4.10b)$$

because the integral of $\mathbf{J}^2$ is proportional to the energy supplied by the source. Note that in (4.10a) the integral is extended onto the bounded support of the test function $\varphi(x, y, z)$. Therefore, on the basis of the assumption $(2d) \ll 2\pi/(\omega\sqrt{\varepsilon_0\mu_0})$, it can be written as follows:

$$\begin{aligned}
<\mathbf{J}(x, y, z), \varphi(x, y, z)> &= \iiint \mathbf{J}(x, y, z)\varphi(x, y, z) dxdydz \\
&= \iint \left\{ \int_{-d}^{d} \mathbf{J}(x, y, z)\varphi(x, y, z)\, dz \right\} dxdy \\
&\approx \iint \left\{ \int_{-d}^{d} \mathbf{J}(x, y, 0)\varphi(x, y, 0)\, dz \right\} dxdy \\
&= \iint \{2d\mathbf{J}(x, y, 0)\varphi(x, y, 0)\} dxdy \\
&= \iiint 2d\mathbf{J}(x, y, 0)\varphi(x, y, z)\delta(z)\, dxdydz \\
&= < 2d\mathbf{J}(x, y, 0)\, \delta(z), \varphi(x, y, z) >. \quad (4.10c)
\end{aligned}$$

From this one concludes that

$$\begin{aligned}
\mathbf{J}(x, y, z) &\approx (2d)\mathbf{J}(x, y, 0)\delta(z) \\
&\equiv \mathbf{J}_0(x, y)\delta(z) \quad (4.11a)
\end{aligned}$$

with

$$\mathbf{J}_0(x,y) = (2d)\mathbf{J}(x,y,0) \approx d[\mathbf{J}(x,y,d-0) + \mathbf{J}(x,y,-d+0)]. \tag{4.11b}$$

Therefore the tangential and normal components of $\mathbf{J}_0(x,y)$ are

$$\mathbf{J}_{0t}(x,y) \approx d[\sigma\mathbf{E}_t(x,y,d-0) + \sigma\mathbf{E}_t(x,y,-d+0)] \approx d\sigma[\mathbf{E}_t^{(1)} + \mathbf{E}_t^{(2)}] \tag{4.11c}$$

and

$$J_{0n}(x,y) \approx d[\sigma_n E_n(x,y,d-0) + \sigma_n E_n(x,y,-d+0)] \approx d\frac{\sigma_n}{\varepsilon_n'}[\varepsilon_1' E_n^{(1)} + \varepsilon_2' E_n^{(2)}]. \tag{4.11d}$$

Here we assume that all the conductivities $\sigma, \sigma_1, \sigma_2$, and σ_n are *finite* and put (see (3.37c,d))

$$\varepsilon_1' = \varepsilon_1 + \frac{i\sigma_1}{\omega}, \qquad \varepsilon_2' = \varepsilon_2 + \frac{i\sigma_2}{\omega}, \qquad \varepsilon_n' = \varepsilon_n + \frac{i\sigma_n}{\omega}. \tag{4.12}$$

Note that in writing (4.11d) we have used the simple boundary relations on the interfaces $z = \pm d$, namely,

$$\varepsilon_1' E_n^{(1)} = \varepsilon_n' E_n(x,y,-d+0), \qquad \varepsilon_2' E_n^{(2)} = \varepsilon_n' E_n(x,y,d-0), \tag{4.13}$$

valid for monochromatic fields (see (3.37a–c)).

By a similar reasoning, we get also

$$\mathbf{P}_{0t}(x,y) \approx d(\varepsilon - \varepsilon_0)[\mathbf{E}_t^{(1)} + \mathbf{E}_t^{(2)}], \tag{4.14a}$$

$$P_{0n}(x,y) \approx d\frac{\varepsilon_n - \varepsilon_0}{\varepsilon_n'}[\varepsilon_1' E_n^{(1)} + \varepsilon_2' E_n^{(2)}], \tag{4.14b}$$

$$\mathbf{M}_{0t}(x,y) \approx d(\mu - \mu_0)[\mathbf{H}_t^{(1)} + \mathbf{H}_t^{(2)}], \tag{4.15a}$$

$$M_{0n}(x,y) \approx d\frac{\mu_n - \mu_0}{\mu_n}[\mu_1 H_n^{(1)} + \mu_2 H_n^{(2)}]. \tag{4.15b}$$

It is obvious that the equations (4.11c)–(4.15b) are nothing but the equations (4.9a–f) with

$$\sigma_1^* = \sigma_2^* = d\sigma, \qquad \sigma_{1n} = d\frac{\sigma_n}{\varepsilon_n'}\varepsilon_1', \qquad \sigma_{2n} = d\frac{\sigma_n}{\varepsilon_n'}\varepsilon_2', \tag{4.15c}$$

$$\chi_1^e = \chi_2^e = d\frac{\varepsilon - \varepsilon_0}{\varepsilon_0}, \qquad \chi_{1n}^e = d\frac{\varepsilon_n - \varepsilon_0}{\varepsilon_n'}\frac{\varepsilon_1'}{\varepsilon_0}, \tag{4.15d}$$

$$\chi_{2n}^e = d\frac{\varepsilon_n - \varepsilon_0}{\varepsilon_n'}\frac{\varepsilon_2'}{\varepsilon_0}, \qquad \chi_1^m = \chi_2^m = d\frac{\mu - \mu_0}{\mu_0}, \tag{4.15e}$$

$$\chi_{1n}^m = d\frac{\mu_n - \mu_0}{\mu_n}\frac{\mu_1}{\mu_0}, \qquad \chi_{2n}^e = d\frac{\mu_n - \mu_0}{\mu_n}\frac{\mu_2}{\mu_0}. \tag{4.15f}$$

By putting (4.11c)–(4.15b) in (4.7a,b), we get finally

$$\mathbf{n} \times (\mathbf{H}_t^{(2)} - \mathbf{H}_t^{(1)}) = e[\mathbf{E}_t^{(1)} + \mathbf{E}_t^{(2)}] + h_{1n}\mathbf{n} \times \text{grad}H_n^{(1)} + h_{2n}\mathbf{n} \times \text{grad}H_n^{(2)} \tag{4.16a}$$

$$\mathbf{n} \times (\mathbf{E}_t^{(2)} - \mathbf{E}_t^{(1)}) = -h[\mathbf{H}_t^{(1)} + \mathbf{H}_t^{(2)}] + e_{1n}\mathbf{n} \times \text{grad}E_n^{(1)} + e_{2n}\mathbf{n} \times \text{grad}E_n^{(2)}, \tag{4.16b}$$

where

$$e = -i\omega d(\varepsilon' - \varepsilon_0), \qquad h = -i\omega d(\mu - \mu_0) \tag{4.17a}$$

$$h_{1n} = -d\frac{\mu_1}{\mu_0}\frac{\mu_n - \mu_0}{\mu_n}, \qquad h_{2n} = -d\frac{\mu_2}{\mu_0}\frac{\mu_n - \mu_0}{\mu_n} \tag{4.17b}$$

$$e_{1n} = -d\frac{\varepsilon_1'}{\varepsilon_0}\frac{\varepsilon_n' - \varepsilon_0}{\varepsilon_n'}, \qquad e_{2n} = -d\frac{\varepsilon_2'}{\varepsilon_0}\frac{\varepsilon_n' - \varepsilon_0}{\varepsilon_n'} \tag{4.17c}$$

with

$$\varepsilon' = \varepsilon + \frac{i\sigma}{\omega}. \tag{4.17d}$$

Equations (4.16a,b) are nothing but the approximate boundary relations on the sheet which simulates the very thin layer shown in Fig. 4.1.

PROBLEMS

1. Let a slab of width ($2d$), filled with a simple *isotropic* material, be illuminated by a linearly polarized monochromatic wave whose electric field is parallel to the slab (see Fig. 1 below).

(a) Write the explicit expressions of the incident wave and boundary conditions to be satisfied by the total wave.

(b) Find an explicit expression of the reflected wave in the region above the slab and derive an expression for the reflection coefficient R. Plot the curves which show the variations of $|R|$ and $\arg R$ with $(2d/\lambda) \in (0, 1)$ for different values of ε, μ, and σ. Here λ stands for the free-space wavelength (choose $\lambda = 1$).

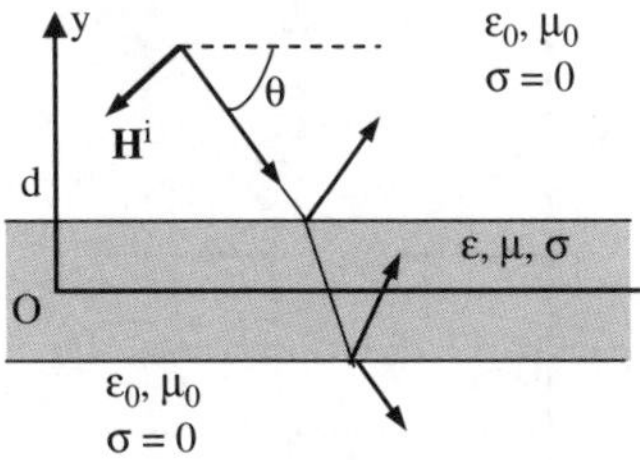

Figure 1

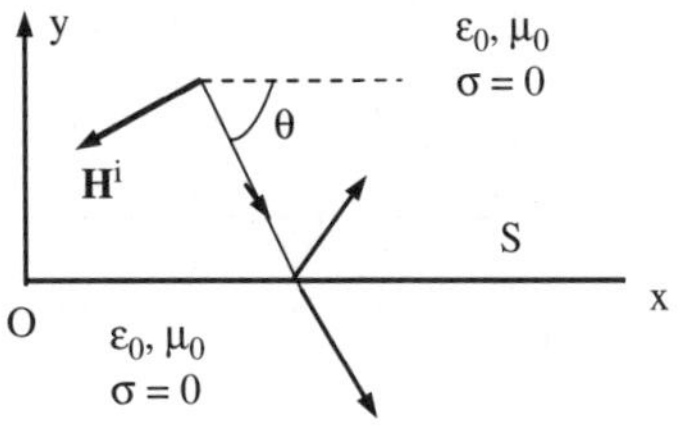

Figure 2

(c) Find an explicit expression of the transmitted wave in the region under the slab and derive an expression for the transmission coefficient T. Plot the curves which show the variations of $|T|$ and $\arg T$ with $(2d/\lambda) \in (0, 1)$ for different values of ε, μ, and σ.

(d) Consider now the material sheet S which models the above-mentioned slab when $(2d) \ll \lambda$ and assume the same incident wave as above (see Fig. 2 above). Write the boundary conditions to be satisfied on S.

(e) Find the explicit expressions of the reflected and transmitted waves generated by the material sheet S mentioned in (d).Derive the explicit expressions of the reflection coefficient R_S and transmission coefficient T_S. Plot the curves mentioned in (b) and (c) for R_S and T_s with the same values of the parameters.

(f) Compare the results obtained in (b), (c), and (e) through the curves showing the variations of $|R/R_S|$, $[\arg R - \arg R_S]$, $|T/T_S|$ and $[\arg T - \arg T_S]$ with $(2d/\lambda)$ for different values of $(\varepsilon, \mu, \sigma)$ and θ. Discuss the particular cases of $\theta = \pi/2$ and $\theta \to 0$.

2. Let the half-spaces $z > d$ and $z < d$ be filled with simple materials and assume that in the region $z < d$ the wave can penetrate until the level $z = -d$ such that $2d = np$ (see Fig. 1 and Fig. 2 below), where p stands for the penetration depth ($p = \sqrt{2/(\omega\mu\sigma)}$) while n is an integer ($n = 1$, 2, or 5).

   (a) Find the constitutive relations pertinent to the material sheet S which models the slab of width 2d (cf. (4.11c)–(4.15b)).

(b) What are the boundary conditions to be satisfied on S?

(c) Solve problem 1 first for the two-part space divided by the interface $z = d$ (see Fig. 1 below) and then for the half-space above the material sheet S mentioned in (b) (see Fig. 3 below). To compare the results, plot the curves representing the variations of $|R/R_S|$ and $[\arg R - \arg R_S]$ with $(2d/\lambda) \in (0, 1)$ for different values of $(\varepsilon, \mu, \sigma)$ and n.

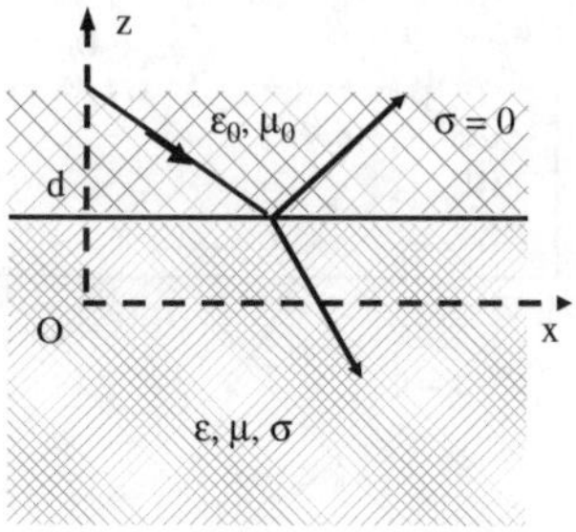

Figure 1

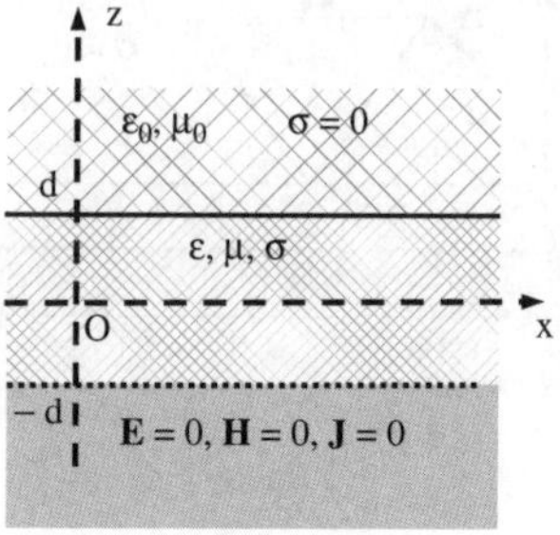

Figure 2

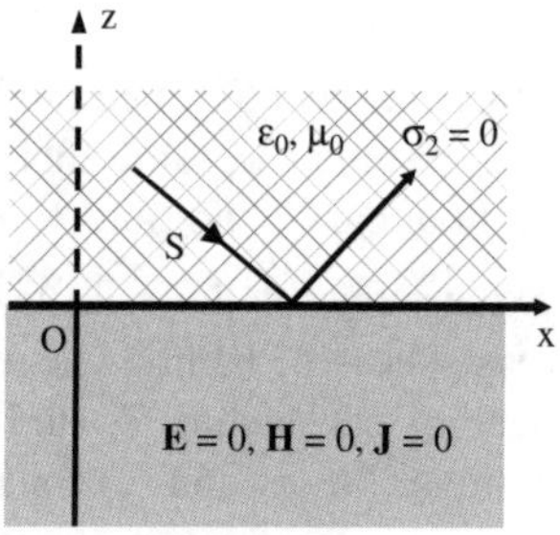

Figure 3

3. Solve problems 1 and 2 for the case when the magnetic field is parallel to the slab (see Fig. 1 and Fig. 2 below).

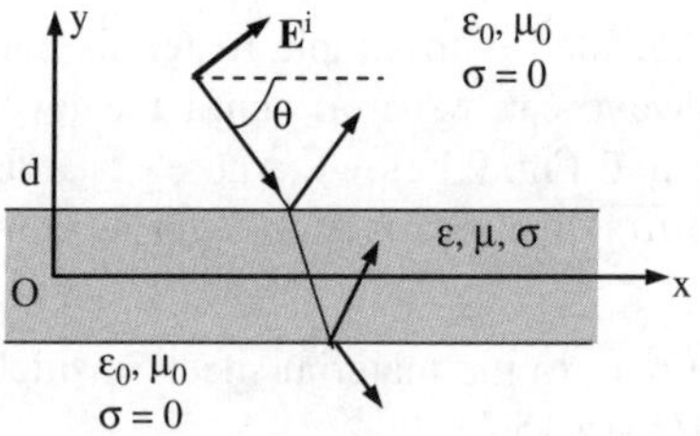

Figure 1

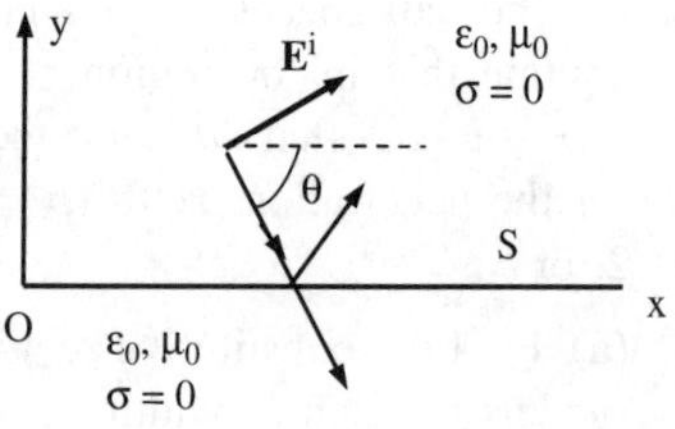

Figure 2

A. Some Extensively Considered Particular Cases

When one has

$$\varepsilon_n = \varepsilon_0, \qquad \sigma_\mathrm{n} = 0, \qquad \mu = \mu_n = \mu_0 \tag{4.18a}$$

while

$$\varepsilon' \neq \varepsilon_0, \tag{4.18b}$$

the boundary conditions (4.16a,b) are reduced to

$$\mathbf{E}_t^{(2)} = \mathbf{E}_t^{(1)}, \qquad \mathbf{n} \times \left(\mathbf{H}_t^{(2)} - \mathbf{H}_t^{(1)}\right) = \frac{1}{R}\mathbf{E}_t^{(1)} \tag{4.19a}$$

with

$$R = \frac{1}{2e} = \frac{\mathrm{i}}{2\omega d[(\varepsilon - \varepsilon_0) + \frac{i\sigma}{\omega}]}. \tag{4.19b}$$

This is the case of *resistive sheet* considered first by Levi-Civita (see [20, p. 19]) and then by many others [21, 22]. As is seen from the conditions (4.18a,b), a resistive sheet simulates a nonmagnetic *anisotropic* thin layer. The following particular cases are especially important from practical points of view:

a. $\quad \sigma = 0, \quad \varepsilon \neq \varepsilon_0 \quad \Rightarrow \quad R = \dfrac{\mathrm{i}}{2\omega d(\varepsilon - \varepsilon_0)}$ (4.20a)

b. $\quad \sigma \neq 0, \quad \varepsilon = \varepsilon_0 \quad \Rightarrow \quad R = \dfrac{1}{2\sigma d}$ (4.20b)

The dual of the situation cited above is met when one has

$$\mu_n = \mu_0, \qquad \sigma = \sigma_n = 0, \qquad \varepsilon = \varepsilon_n = \varepsilon_0, \qquad \mu \neq \mu_0. \tag{4.21}$$

In such a case, (4.16a,b) are reduced to

$$\mathbf{H}_t^{(2)} = \mathbf{H}_t^{(1)}, \qquad \mathbf{n} \times (\mathbf{E}_t^{(2)} - \mathbf{E}_t^{(1)}) = \frac{1}{G}\mathbf{H}_t^{(1)} \tag{4.22a}$$

with

$$G = -\frac{1}{2h} = \frac{-i}{2\omega d(\mu - \mu_0)}. \tag{4.22b}$$

By analogy with the resistive sheet considered above, a sheet whose boundary properties are described by (4.22a) was called a *magnetically conducting* sheet* by certain authors [20, 23].

For a nonmagnetic slab, one has $h = h_{1n} = h_{2n} = 0$ (see (4.17a,b)) and, hence,

$$\mathbf{n} \times (\mathbf{H}_t^{(2)} - \mathbf{H}_t^{(1)}) = \frac{1}{2R}[\mathbf{E}_t^{(1)} + \mathbf{E}_t^{(2)}] \tag{4.23a}$$

$$\mathbf{n} \times (\mathbf{E}_t^{(2)} - \mathbf{E}_t^{(1)}) = e_{1n}\mathbf{n} \times \text{grad}E_n^{(1)} + e_{2n}\mathbf{n} \times \text{grad}E_n^{(2)} \tag{4.23b}$$

with R and $e_{1,2n}$ given by (4.19b) and (4.17c), respectively. It is worthwhile to draw attention to the fact that the sheet with properties (4.23a,b) *can simulate an isotropic layer too* whereas a resistive sheet corresponds only to an anisotropic slab. The *combined sheet* considered in Senior [23] is a special case of (4.23a,b), where $e_{1n} = e_{2n}$..

B. A Remark

If the slab is composed of several simple layers of the form shown in Fig. 4.2, then the coefficients e (or h) to be associated with $\mathbf{E}_t^{(1)}$ and $\mathbf{E}_t^{(2)}$ (or $\mathbf{H}_t^{(1)}$ and $\mathbf{H}_t^{(2)}$) become different. In such a case, one has (4.9a–f) instead of (4.11c)–(4.15b) (*Determination of the constitutive coefficients in (4.9a–f) constitutes still an open problem*). Similarly, when the constitutive parameters of the slab depend on z, the result is quite similar to this case.

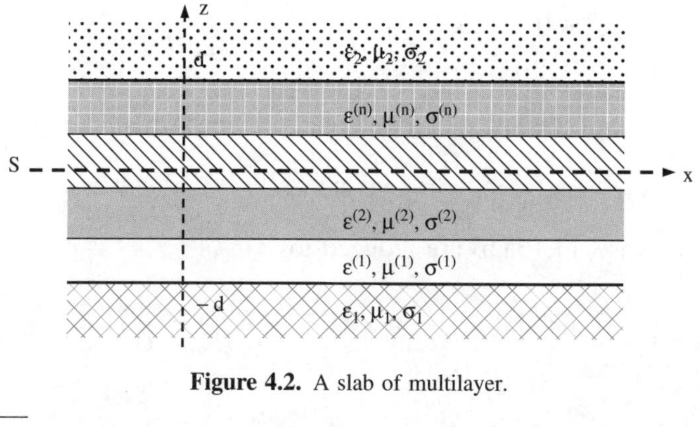

Figure 4.2. A slab of multilayer.

*It goes without saying that such a name is not convenient because it invokes hypothetical magnetic conductance, which is meaningless in real situations whereas (4.22a, b) is quite meaningful.

C. Equivalent Boundary Conditions for Monochromatic Fields

By substituting (4.9a–f) in (4.7a,b) we find, for a monochromatic wave of frequency ω, the boundary relations on a simple planar material sheet in their explicit forms, namely,

$$\begin{aligned}\mathbf{n}\times\mathbf{H}_t^{(1)} - \mathbf{n}\times\mathbf{H}_t^{(2)} + e_1\mathbf{E}_t^{(1)} + e_2\mathbf{E}_t^{(2)} + h_{1n}\mathbf{n}\times\mathrm{grad}H_n^{(1)} \\ + h_{2n}\mathbf{n}\times\mathrm{grad}H_n^{(2)} = 0\end{aligned} \tag{4.24a}$$

$$\begin{aligned}h_1\mathbf{n}\times\mathbf{H}_t^{(1)} + h_2\mathbf{n}\times\mathbf{H}_t^{(2)} + \mathbf{E}_t^{(1)} - \mathbf{E}_t^{(2)} + e_{1n}\mathrm{grad}E_n^{(1)} \\ + e_{2n}\mathrm{grad}E_n^{(2)} = 0\end{aligned} \tag{4.24b}$$

with

$$e_\nu = \sigma^*{}_\nu - i\omega\varepsilon_0\chi^e_\nu, \qquad \nu = 1,2 \tag{4.25a}$$

$$h_\nu = -i\omega\mu_0\chi^m{}_\nu, \tag{4.25b}$$

$$e_{\nu n} = [\sigma_{\nu n} - i\omega\varepsilon_0\chi^e{}_{\nu n}]/(i\omega\varepsilon_0)], \tag{4.25c}$$

$$h_{\nu n} = -\chi^m{}_{\nu n}. \tag{4.25d}$$

Note that (4.24b) was obtained through cross-multiplication of (4.7b) with **n**, and the remark made in (B) above was taken into account.

The relations (4.24a,b) are, in general, the sufficient couple of boundary conditions to be considered in solving time-harmonic problems. In practice, however, the boundary conditions are not always given in these original forms as (4.24a,b) but rather as particular linear combinations of them. In such a case they are of the following form:

$$\begin{aligned}a_1\mathbf{n}\times\mathbf{H}_t^{(1)} + a_2\mathbf{n}\times\mathbf{H}_t^{(2)} + a_3\mathbf{E}_t^{(1)} + a_4\mathbf{E}_t^{(2)} + a_5\mathbf{n}\times\mathrm{grad}H_n^{(1)} \\ + a_6\mathbf{n}\times\mathrm{grad}H_n^{(2)} + a_7\mathrm{grad}E_n^{(1)} + a_8\mathrm{grad}E_n^{(2)} = 0,\end{aligned} \tag{4.26a}$$

$$\begin{aligned}b_1\mathbf{n}\times\mathbf{H}_t^{(1)} + b_2\mathbf{n}\times\mathbf{H}_t^{(2)} + b_3\mathbf{E}_t^{(1)} + b_4\mathbf{E}_t^{(2)} + b_5\mathbf{n}\times\mathrm{grad}H_n^{(1)} \\ + b_6\mathbf{n}\times\mathrm{grad}H_n^{(2)} + b_7\mathrm{grad}E_n^{(1)} + b_8\mathrm{grad}E_n^{(2)} = 0.\end{aligned} \tag{4.26b}$$

The table C formed by a_j and b_j, namely

$$C = \begin{bmatrix} a_1 & a_2 & \dots & a_8 \\ b_1 & b_2 & \dots & b_8 \end{bmatrix}, \tag{4.27}$$

is called a *boundary relation table* of the simple planar sheet [16, 19]. The table correspondding to the original form (4.24a,b) is, for example, as follows:

$$C = \begin{bmatrix} 1 & -1 & e_1 & e_2 & h_{1n} & h_{2n} & 0 & 0 \\ h_1 & h_2 & 1 & -1 & 0 & 0 & e_{1n} & e_{2n} \end{bmatrix}. \tag{4.28}$$

It is obvious that any table written as (4.27) cannot always be treated as a boundary relation table connected with a certain simple material planar sheet located on the interface of two simple media. The necessary and sufficient condition for C to represent a couple of boundary relations can be found by simple algebraic manipulations. They can be summarized in the form of three theorems given below [16]. In what follows we use the following abbreviations:

$$\Delta^i_j = a_i b_j - b_i a_j = -\Delta^j{}_i \tag{4.29a}$$

$$\Delta^{ij}_k = (a_i + a_j)b_k - (b_i + b_j)a_k = -\Delta^k{}_{ij} \tag{4.29b}$$

$$\Delta^i{}_{jk} = a_i(b_j + b_k) - b_i(a_j + a_k) = -\Delta^{jk}{}_i \tag{4.30a}$$

and

$$\Delta^{ij}{}_{km} = (a_i + a_j)(b_k + b_m) - (b_i + b_j)(a_k + a_m) = -\Delta^{km}{}_{ij} \tag{4.30b}$$

THEOREM 4.1 **(Necessary Conditions)**

Let C represent a couple of boundary relations on a simple planar material sheet. Then one has

$$\text{Rank} \begin{bmatrix} a_1 + a_2 & a_7 & a_8 \\ b_1 + b_2 & b_7 & b_8 \end{bmatrix} \leq 1 \tag{4.31}$$

$$\text{Rank} \begin{bmatrix} a_3 + a_4 & a_5 & a_6 \\ b_3 + b_4 & b_5 & b_6 \end{bmatrix} \leq 1. \tag{4.32}$$

THEOREM 4.2 **(Sufficient Conditions)**

If the relations cited in Theorem 4.1 are fulfilled and $\Delta^1{}_2$ and $\Delta^3{}_4$ are different from zero, then C represents a couple of boundary relations on a uniquely determined simple planar material sheet.

THEOREM 4.3

If C represents a couple of boundary relations on a simple material sheet and all the sums $a_j + a_{j+1}$ and $b_j + b_{j+1}$ $(j = 1, 3)$ are zero, then the sheet is uniquely determined and one has

$$\Delta^1{}_2 = \Delta^3{}_4 = \Delta^5{}_6 = \Delta^7{}_8 = 0 \tag{4.33a}$$

and

$$\Delta^1{}_7 \neq 0 \text{ (or } \Delta^1{}_8 \neq 0), \qquad \Delta^3{}_5 \neq 0 \text{ (or } \Delta^3{}_6 \neq 0) \tag{4.33b}$$

if one of the coefficients (a_7, b_7) (or (a_8, b_8)) and one of (a_5, b_5) (or (a_6, b_6)) are different from zero.

In order to grasp the meaning of these theorems, it will be useful to see their proofs. To this end, consider first Theorem 4.1 and assume that C represents a boundary conditions couple. That means that the equations (4.26a,b) are obtained from (4.24a,b) through certain linear transformations. Hence, inversely, (4.26a,b) can (and must) be reduced to (4.24a,b) by suitable linear transformations. In other words, there must be a couple of constants (A, B) and (A', B') such that

$$Aa_1 + Bb_1 = 1, \qquad Aa_2 + Bb_2 = -1 \tag{4.34a}$$

$$Aa_7 + Bb_7 = 0, \qquad Aa_8 + Bb_8 = 0 \tag{4.34b}$$

$$Aa_3 + Bb_3 = \mathrm{e}_1, \qquad Aa_4 + Bb_4 = \mathrm{e}_2 \tag{4.34c}$$

$$Aa_5 + Bb_5 = \mathrm{h}_{1n}, \qquad Aa_6 + Bb_6 = h_{2n} \tag{4.34d}$$

$$A'a_3 + B'b_3 = 1, \qquad A'a_4 + B'b_4 = -1 \tag{4.35a}$$

$$A'a_5 + B'b_5 = 0, \qquad A'a_6 + B'b_6 = 0 \tag{4.35b}$$

$$A'a_1 + B'b_1 = h_1, \qquad A'a_2 + B'b_2 = h_2 \tag{4.35c}$$

$$A'a_7 + B'b_7 = e_{1n}, \qquad A'a_8 + B'b_8 = e_{2n}. \tag{4.35d}$$

Note that the first equations in (4.34a,b) and those in (4.35a,b) enable us to determine the constants $(A,\ B)$ and (A', B') while the remaining ones are used to determine the constitutive constants $e_1, e_2, h_{1n}, h_{2n}, h_1, h_2, e_{1n}$, and e_{2n}. Notice also that the second equations in (4.34a) and (4.35a) can be replaced by homogeneous equations

$$A(a_1 + a_2) + B(b_1 + b_2) = 0, \qquad A'(a_3 + a_4) + B'(b_3 + b_4) = 0. \tag{4.36}$$

These equations are obtained by the sum of two equations in (4.34a) and (4.35a).

The compatibility condition of the homogeneous equations in (4.34b) and (4.36) gives (4.31) mentioned in Theorem 4.1. If (4.31) is satisfied, then the constants A and B can be determined through the first equations in (4.34a) and (4.36).

The system in (4.35a,b) is obtained from (4.34a,b) by the following substitutions of indices:

$$1 \leftrightarrow 3, \qquad 2 \leftrightarrow 4, \qquad 7 \leftrightarrow 5, \qquad 8 \leftrightarrow 6 \tag{4.37}$$

Since with these substitutions (4.31) is reduced to (4.32), the proof of Theorem 4.1 becomes complete.

Let us now consider Theorem 4.2. Since $\Delta^1{}_2 \neq 0, A$ and B are solved uniquely from the first equations in (4.34a) and (4.36) as follows:

$$A = \frac{b_1 + b_2}{\Delta_2^1}, \qquad B = -\frac{a_1 + a_2}{\Delta_2^1}. \tag{4.38a}$$

By putting these expressions of A and B into (4.34 c,d), one gets the expressions of e_1, e_2, h_{1n}, and h_{2n} as follows:

$$e_1 = \frac{\Delta_{12}^3}{\Delta_2^1}, \qquad e_2 = \frac{\Delta_{12}^4}{\Delta_2^1}, \qquad h_{1n} = \frac{\Delta_{12}^5}{\Delta_2^1}, \qquad h_{2n} = \frac{\Delta_{12}^6}{\Delta_2^1}. \tag{4.38b}$$

As to the system (4.35a–d), its solution can be obtained from (4.38a,b) through the substitutions of indices given by (4.37) and

$$A \to A', \quad B \to B', \tag{4.39}$$

$$e_1 \leftrightarrow h_1, \quad e_2 \leftrightarrow h_2, \quad e_1 \leftrightarrow h_1, \quad h_{1n} \leftrightarrow e_{1n}, \quad h_{2n} \leftrightarrow e_{2n}. \tag{4.40}$$

Thus one gets (if $\Delta^3{}_4 \neq 0$),

$$A' = \frac{b_3 + b_4}{\Delta_4^3}, \qquad B' = -\frac{a_3 + a_4}{\Delta_4^3}, \tag{4.41}$$

$$h_1 = \frac{\Delta_{34}^1}{\Delta_4^3}, \qquad h_2 = \frac{\Delta_{34}^2}{\Delta_4^3}, \qquad e_{1n} = \frac{\Delta_{34}^7}{\Delta_4^3}, \qquad e_{2n} = \frac{\Delta_{34}^8}{\Delta_4^3}. \tag{4.42}$$

Thus the proof of Theorem 4.2 is over.

Finally, let us consider the case mentioned in Theorem 4.3. In this case the relations $a_1 + a_2 = 0$ and $b_1 + b_2 = 0$ yield $\Delta^1{}_2 = 0$, which means that the second equation in (4.34a) linearly depends on the first one and is to be discarded from the list. Similarly, the homogeneous equations in (4.34b) give $\Delta^7{}_8 = 0$, which shows that one of them is to be omitted. Assume that a_7 (or b_7) is different from zero and omit the second equation in (4.34b). Hence, the remaining equations, namely the first equations in (4.34a) and (4.34b), has to be used to determine A and B. This requires the relation $\Delta^1{}_7 \neq 0$ and gives the unique solution:

$$A = \frac{b_7}{\Delta_7^1}, \qquad B = -\frac{a_7}{\Delta_7^1}, \qquad e_1 = \frac{\Delta_7^3}{\Delta_7^1}, \tag{4.43a}$$

$$e_2 = \frac{\Delta_7^4}{\Delta_7^1}, \qquad h_{1n} = \frac{\Delta_7^5}{\Delta_7^1}, \qquad h_{2n} = \frac{\Delta_7^6}{\Delta_7^1}. \tag{4.43b}$$

If both a_7 and b_7 are equal to zero, then the solution is obtained by replacing all indices 7 by 8.

By a similar reasoning one gets from the equations in (4.35a–d) that if one of the coefficients a_5 and b_5 differs from zero, then $\Delta^3{}_4 = 0, \Delta^5{}_6 = 0$ while $\Delta^3{}_5 \neq 0$ and

$$A' = \frac{b_5}{\Delta_5^3}, \qquad B' = -\frac{a_5}{\Delta_5^3}, \qquad h_1 = \frac{\Delta_5^1}{\Delta_5^3}, \tag{4.44a}$$

$$h_2 = \frac{\Delta_5^2}{\Delta_5^3}, \qquad e_{1n} = \frac{\Delta_5^7}{\Delta_5^3}, \qquad e_{2n} = \frac{\Delta_5^8}{\Delta_5^3}. \tag{4.44b}$$

If one has $a_5 = b_5 = 0$, then all indices 5 appearing in (4.43a,b) are to be replaced by 6. Note that the expressions given by (4.44a,b) can also be obtained from (4.43a,b) by the substitutions (4.37), (4.39), and (4.40).

Thus the proofs of Theorems 4.1–4.3 are complete.

D. An Example. Resistive Conditions

Let us consider the table

$$C = \begin{bmatrix} 1 & -1 & \gamma & (1/R - \gamma) & 0 & 0 & 0 & 0 \\ 0 & 0 & 1 & -1 & 0 & 0 & 0 & 0 \end{bmatrix}, \tag{4.45}$$

where R denotes a nonzero, finite complex number while γ is an arbitrary complex parameter. It is evident that for all values of γ this table corresponds to the *resistive sheet* considered in Section 4.3.1.A.

As we can easily check, the relations cited in Theorem 4.1 are fulfilled while neither the relations cited in Theorem 4.2 nor those in Theorem 4.3 are met. Hence the sheet, if it exists, is *not uniquely* determined. From (4.34a)–(4.35d) one gets

$$A = 1, \qquad \text{B} = \text{arbitrary}, \qquad A' = 0, \qquad B' = 1, \tag{4.46}$$

$$h_1 = h_2 = h_{1n} = h_{2n} = 0, \qquad e_{1n} = e_{2n} = 0, \tag{4.47}$$

and

$$e_1 = \gamma + B, \qquad e_2 = (1/R - \gamma) - B, \tag{4.48}$$

which yields also

$$e_1 + e_2 = 1/R. \tag{4.49}$$

The nonuniqueness of the sheet proceeds from (4.48), which involves an undetermined constant B. Hence e_1 and e_2 cannot be determined uniquely. If one supposes that the sheet simulates a slab similar to that considered in Section 4.3.1.A, then (4.47) and (4.49) yield (4.18a) and (4.18b), respectively. That means first of all that the conditions (4.18a) are necessary for a slab similar to that shown in Fig. 4.1 to be simulated by a resistive sheet. This requires, in particular, $\varepsilon \neq \varepsilon_n = \varepsilon_0$ or/and $\sigma \neq \sigma_n = 0$, which claim that the material of the slab is always anisotropic. When R is real and positive, (4.48) or, equivalently, (4.19b) yields $\varepsilon = \varepsilon_n = \varepsilon_0$ and $\sigma = 1/(2Rd)$, which is merely (4.20b). In such a case the anisotropy of the material is only in its conductivity. From this we conclude that a thin slab composed of *isotropic* conducting material *cannot*, in general, be simulated by a resistive sheet. However, if the normal component of the total electric field inside the slab is zero, then the violation of the condition $\sigma_n = 0$ has no effect on the accuracy of the simulation by resistive sheet because σ_n affects the field only through the product $(\sigma_n E_n)$. Note that the poor quality of the approximation to be obtained by resistive sheet simulation when $E_n \neq 0$ is first observed in Senior and Volakis [24] without discussing the fact lying behind the phenomenon.

E. Critical Frequencies and Boundary Conditions of Impedance Type

The study of the electromagnetic wave propagation in a region *outside* certain bodies of differrent physical properties requires, in general, the determination of the wave inside the bodies also, which may be beyond the scope of the study. The knowledge of the wave inside the bodies is needed to ensure the boundary conditions on their surfaces. If, however, the boundary conditions are such that only the external field components on the surfaces of the bodies are interrelated by them, then the determination of the field inside the bodies becomes avoided, which greatly simplifies the task. The simple condition $\mathbf{E}_t^{(2)} = 0$ written on the surface of a perfectly conducting body is the classical example for such type of conditions. The superscript (2) in

$\mathbf{E}_t^{(2)}$ refers to the outside region. A rather simple generalization of this condition is

$$\mathbf{E}_t^{(2)} = Z_2 \mathbf{n} \times \mathbf{H}_t^{(2)}, \tag{4.50}$$

where $\mathbf{n}$ stands for the outward unit normal on the surface and $Z_2 \neq 0$ is a given complex scalar. Such a relation, which is termed *impedance type boundary condition*, has been extensively used in analyzing the radio wave propagation over the Earth (for early applications see, for example, references [25–27]). The same condition was also used to represent conducting layers (when the thickness is much greater than the penetration depth but much smaller than the free-space wavelength), metal-backed layers (when $|\varepsilon\mu| \gg \varepsilon_0\mu_0$), and corrugated surfaces (when the spacing of the grooves exceeds their individual width and there are many corrugations per wavelength). Nevertheless, in spite of the facilities resulted from the use of them, the conditions of impedance type were always regarded as to be doubtful because the expression of the surface impedance—that is, Z_2 in (4.50)—depends, in general, on the source configurations. Among the investigations devoted to the study of the validity of impedance-type conditions, we can mention references [28,29,30,31]. It should be noted that in all of these investigations except Idemen [28], attention was focused only on the incidence angle. But, as we shall see below, the effect of the frequency on the validity of impedance type conditions is of vital importance (see also Idemen [28]). Our aim is now to discuss the surface impedance concept by viewing the surface of discontinuity as a physical object having its own constitutive equations.

Since the condition (4.50) is different from what are satisfied on the interface between two media, it must be considered to be related to a material sheet located on the interface. In what follows we will assume that the sheet is simple and its constitutive parameters are given by (4.9.a–f). Then, as a result of (4.50), the first row of the boundary relation table related to the sheet is as follows:

$$[0 \quad Z_2 \quad 0 \quad -1 \quad 0 \quad 0 \quad 0 \quad 0]. \tag{4.51}$$

Since $Z_2 \neq 0$ by assumption, both $a_1 + a_2$ and $a_3 + a_4$ differ from zero, which shows that the situation is beyond Theorem 4.3. Hence the requirements of Theorem 4.2 must be met. From $\Delta_2^1 \neq 0$ and

$\Delta_4^3 \neq 0$ one gets first

$$b_1 \neq 0, \qquad b_3 \neq 0. \tag{4.52a}$$

Then the requirements

$$\text{Rank}\begin{bmatrix} Z_2 & 0 & 0 \\ b_1+b_2 & b_7 & b_8 \end{bmatrix} \leq 1, \qquad \text{Rank}\begin{bmatrix} -1 & 0 & 0 \\ b_3+b_4 & b_5 & b_6 \end{bmatrix} \leq 1$$

yield

$$b_5 = b_6 = b_7 = b_8 = 0. \tag{4.52b}$$

Thus (4.26b) is reduced into

$$b_1\mathbf{n} \times \mathbf{H}^{(1)} + b_2\mathbf{n} \times \mathbf{H}^{(2)} + b_3\mathbf{E}_t^{(1)} + b_4\mathbf{E}_t^{(2)} = 0. \tag{4.52c}$$

As an immediate consequence of the second theorem is that the sheet exists and unique.

From (4.38b) and (4.42) we write

$$e_{1n} = e_{2n} = h_{1n} = h_{2n} = 0 \tag{4.53a}$$

$$e_1 = b_3/b_1, \; h_1 = b_1/b_3 \tag{4.53b}$$

$$e_2 = b_4/b_1 + (b_1+b_2)/(b_1 Z_2) \tag{4.53c}$$

$$h_2 = b_2/b_3 + (b_3+b_4)Z_2/b_3. \tag{4.53d}$$

These mean that if $b_1(\neq 0)$, b_2, $b_3(\neq 0)$, b_4, and $Z_2(\neq 0)$ as well as the frequency ω are given, the constitutive parameters of the sheet can uniquely be determined through the relations (4.53a–d) and (4.25a–d). Hence *there is no problem to discuss*. Note that the properties of the material forming the sheet are different in normal and tangential directions.

Now consider the *converse situation* where the constitutive parameters of a simple sheet are known and conditions under which the sheet seems from one side as an impedance boundary are asked. In this case the problem consists of finding both the complex impedance Z_2 satisfying (4.50) and the complex constants $b_1, b_2, \ldots$ satisfying (4.52a,b) and (4.53b–d). As we shall see now, the solution of this problem does not always exist.

From (4.53b) one concludes first

$$e_1 h_1 = 1, \tag{4.54a}$$

which necessitates both (see (4.25a–d))

$$\chi_1^m \sigma^*{}_1 = 0, \qquad \chi_1^e \chi_1^m < 0 \tag{4.54b,c}$$

and

$$\omega = \sqrt{\frac{-1}{\varepsilon_0 \mu_0 \chi_1^e \chi_1^m}} \equiv \omega_{c_1}. \tag{4.54d}$$

Hence, a given simple material sheet can be seen as an impedance boundary from one side at a certain well-defined *critical frequency* given by (4.54d) if the conditions (4.53a), (4.54b) and (4.54c) are satisfied. If these conditions are met, then the apparent impedance is given by (cf. [28])

$$Z_2 = \frac{h_1 + h_2}{1 + h_1 e_2}. \tag{4.55}$$

As to the other boundary condition to be satisfied on the sheet, it is given by (4.52c) and is not, in general, of impedance type. This means that the sheet acts as a shield for one region only: It protects this region from the effects of the sources lying in the other region. When, in addition to the above-mentioned conditions, the relation

$$b_2 + Z_2 b_4 = 0 \tag{4.56a}$$

is also satisfied, then (4.52c) is reduced to $b_1 \mathbf{n} \times \mathbf{H}^{(1)} + b_3 \mathbf{E}_t^{(1)} = 0$ or

$$\mathbf{E}_t^{(1)} = -Z_1 \mathbf{n} \times \mathbf{H}_t^{(1)} \tag{4.56b}$$

with

$$Z_1 = b_1/b_3 = h_1 = 1/e_1. \tag{4.56c}$$

In such a case we get from (4.53a–d) and (4.56a)

$$Z_2 = h_2 = 1/e_2, \tag{4.57}$$

which yields, similar to (4.54a–d),

$$e_2 h_2 = 1, \tag{4.58a}$$

which necessitates both

$$\chi_2^m \sigma^*{}_2 = 0, \qquad \chi_2^e \chi_2^m < 0 \tag{4.58b,c}$$

and

$$\omega = \sqrt{\frac{-1}{\varepsilon_0 \mu_0 \chi_2^e \chi_2^m}} \equiv \omega_{c2}. \tag{4.58d}$$

This shows that for a surface with different face impedances Z_1 and Z_2 the constitutive parameters satisfy the identity $\omega_{c_1} = \omega_{c_2}$ or, rather explicitly,

$$\chi_1^e \chi_1^m = \chi_2^e \chi_2^m. \tag{4.59a}$$

From (4.54c) and (4.58c) one concludes

$$\sigma^*_{1,2} = 0, \qquad \chi_{1,2}^m < 0, \tag{4.59b}$$

which shows that the material of the sheet is *nonconductive and diamagnetic*.

F. An Illustrative Example [19, 28]

In order to clarify the role played by the frequency in the validity of (4.50), we want to consider the reflection of a plane wave from both the anisotropic slab shown in Fig. 4.3 and the equivalent sheet of impedance boundary shown in Fig. 4.4. The constitutive parameters of the sheet are given through (4.17a–c) while the surface impedance Z_2 is given by (4.55). The explicit expression of the latter is independent of the frequency, namely:

$$Z_2 = -i\omega_c d(\mu - \mu_0) = i\sqrt{\frac{\mu_0}{\varepsilon_0}}\sqrt{\frac{1-\mu/\mu_0}{\varepsilon/\varepsilon_0 - 1}}. \tag{4.60a}$$

The incident field is supposed to be

$$\mathbf{E}^{\text{inc}} = e^{ik_0(y\sin\theta - z\cos\theta)}\mathbf{e}_x, \qquad k_0 = \omega\sqrt{\varepsilon_0\mu_0}, \tag{4.60b}$$

while the other parameters are as follows:

$$d = 10^{-4}, \quad \varepsilon_n = \varepsilon_0, \quad \mu_n = \mu_0, \quad \varepsilon/\varepsilon_0 = 80, \quad \mu/\mu_0 = 0,1. \tag{4.60c}$$

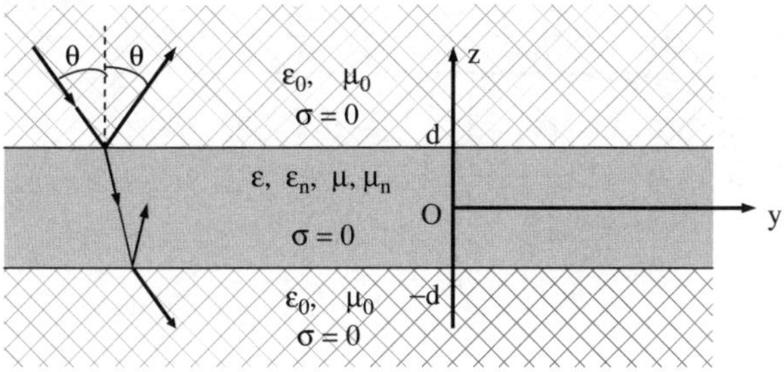

Figure 4.3. Reflection by an anisotropic layer.

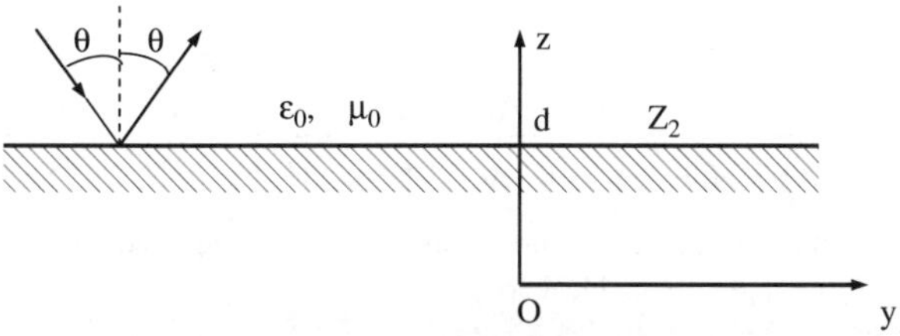

Figure 4.4. Reflection by an impedance boundary.

The reflection coefficients, say R_L and R_S, related to these problems can be found very easily by classical analysis (see problem 1 below). Figures 4.5 and 4.6 show the variations of the amplitude and phase of the ratio (R_L/R_S) as functions of the frequency. The exact value of the critical frequency, computed through (4.58d), namely

$$\omega_c = \frac{1}{d}\frac{1}{\sqrt{\varepsilon_0\mu_0(\varepsilon/\varepsilon_0 - 1)(1 - \mu/\mu_0)}}, \tag{4.60d}$$

is equal to 36×10^{10}. From the curves in Figs. 4.5 and 4.6 it is obviously seen that only at the critical frequency (or near this point) both the amplitudes and the phases of the reflection coefficients are equal to each other. For frequencies lower than the critical one, the disagreement in both the amplitudes and the phases are highly remarkable, whereas for frequencies higher than the critical frequency, only a disagreement in the phases is observable.

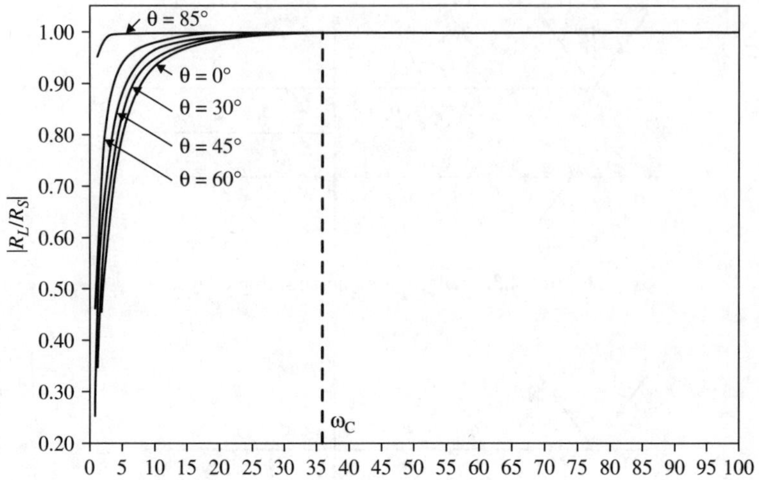

Figure 4.5. Variation of the ratio of amplitudes of the reflection coefficients with frequency for different angles of incidence.

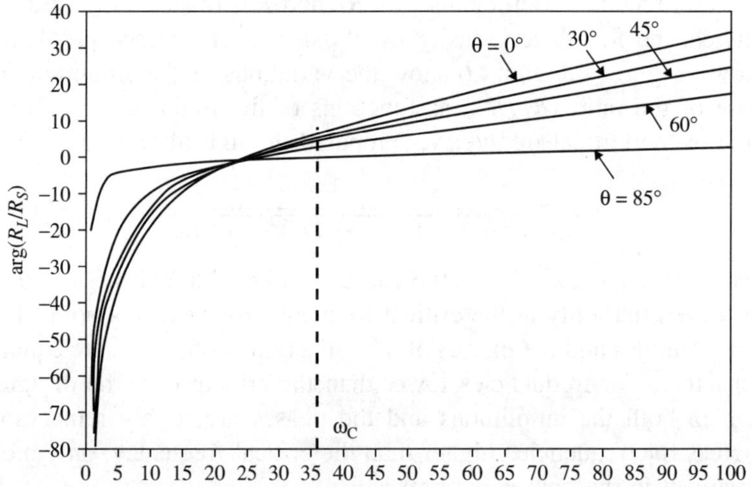

Figure 4.6. Variation of the difference of phases of the reflection coefficients with frequency for different angles of incidence.

PROBLEMS

1. **(a)** Find the reflection coefficients R_L and R_S connected with the configurations shown in Figs. 4.3 and 4.4 for the values of parameters given by (4.60c).

   (b) Solve the same problem for different values of $d, \varepsilon/\varepsilon_0$ and μ/μ_0.

2. Solve problem 1 for the case when $\mathbf{H}^{\text{inc}} = e^{ik_0(y\sin\theta - z\cos\theta)}\mathbf{e}_x$, and $k_0 = \omega\sqrt{\varepsilon_0\mu_0}$.

4.3.2 Cylindrically or Spherically Curved Material Sheet Located Between Two Simple Media

Suppose now that S consists of an infinitely long circular cylinder of radius a or a sphere of radius a. In such a case it is better to formulate the problem in cylindrical polar coordinates (r, ϕ, z) or spherical polar coordinates (r, θ, ϕ), respectively. These coordinates are defined through the well-known usual equations, namely:

in cylindrical coordinates

$$x = r\cos\phi, \quad y = r\sin\phi, \quad z = z$$

with

$$r \in [0, \infty), \quad \phi \in [0, 2\pi), \quad z \in (-\infty, \infty),$$

in spherical coordinates

$$x = r\sin\theta\cos\phi, \quad y = r\sin\theta\sin\phi, \quad z = r\cos\theta,$$

with

$$r \in [0, \infty), \qquad \theta \in [0, \pi], \qquad \phi \in [0, 2\pi).$$

In both cases the function $w(x, y, z)$, the unit normal vector $\mathbf{n}$ and the modulus λ mentioned in Sec. 4.3 are given by

$$w = r - a, \qquad \mathbf{n} = \mathbf{e}_r, \qquad \lambda = 1. \tag{4.61}$$

If S models a cylindrically or spherically curved layer of width $(2d)$ filled with an anisotropic material mentioned in Section 4.1 above, then all the relations and results (i.e., Theorems 4.1–4.3; critical frequencies for impedance type conditions, etc.) derived in Section 4.3.1 for the planar case become valid also for the corresponding curved sheets, provided that necessary changes connected with coordinates are made. For example, (4.8a,b) are replaced by (see Fig. 4.7)

$$[[\mathbf{n} \times \mathbf{E}]] = \mathbf{e}_r \times \left[\mathbf{E}^{(2)} - \mathbf{E}^{(1)}\right] \tag{4.62a}$$

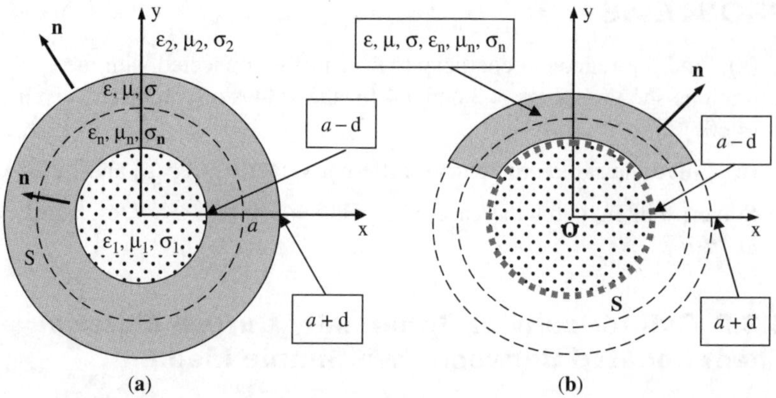

Figure 4.7. A cylindrically or spherically curved layer filled with anisotropic material. (**a**) Complete cylinder or sphere. (**b**) Section of cylindrical or spherical layer.

with

$$\mathbf{E}^{(1)} = \mathbf{E}(a - d - 0, \phi, z), \qquad \mathbf{E}^{(2)} = \mathbf{E}(a + d + 0, \phi, \mathrm{z}) \quad (4.62b)$$

for cylinder while

$$\mathbf{E}^{(1)} = \mathbf{E}(a - d - 0, \theta, \phi), \qquad \mathbf{E}^{(2)} = \mathbf{E}(a + d + 0, \theta, \phi) \quad (4.62c)$$

for sphere. Similarly, (4.11a,b) become

$$\mathbf{J}(r, \phi, z) = \mathbf{J}_0(\phi, z)\delta(r - a) \quad (4.63a)$$

with

$$\mathbf{J}_0(\phi, z) = d[\mathbf{J}(a + d - 0, \phi, \mathrm{z}) + \mathbf{J}(a - d + 0, \phi, z)] \quad (4.63b)$$

and

$$\mathbf{J}(r, \theta, \phi) = \mathbf{J}_0(\theta, \phi)\delta(\mathrm{r} - a) \quad (4.64a)$$

with

$$\mathbf{J}_0(\theta, \phi) = d[\mathbf{J}(a + d - 0, \theta, \phi) + \mathbf{J}(a - d + 0, \theta, \phi)] \quad (4.64b)$$

for the cylindrical and spherical cases, respectively.

It is worthwhile to remark that the above-mentioned similarity that exists between the planar, cylindrical, and spherical cases proceeds from the identity $\lambda = 1$ satisfied for all three cases. As we shall see in the next section, when S consists of a cone one will have $\lambda \neq 1$, which causes substantial differences.

4.3.3 Conical Material Sheet Located Between Two Simple Media

Suppose now that S consists of a rotationally symmetric conical surface defined through $\theta = \theta_0 =$ constant (see Fig. 4.8). Here θ stands for the polar angle measured from the axis of the cone. If S does not consist of an ordinary interface between two simple media and models a thin simple layer about the cone $\theta = \theta_0$, the geometrical structure of the layer is inevitably different near the tip than in a region far away from the tip (see Figs. 4.9 and 4.10). In what follows we will consider these cases separately.

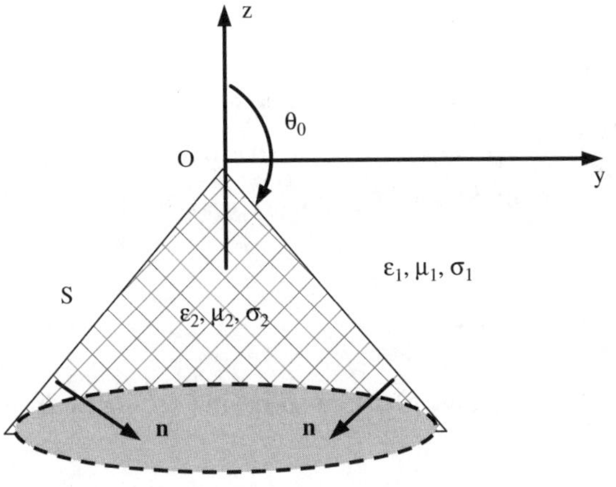

Figure 4.8. A rotationally symmetric conical boundary between two simple media.

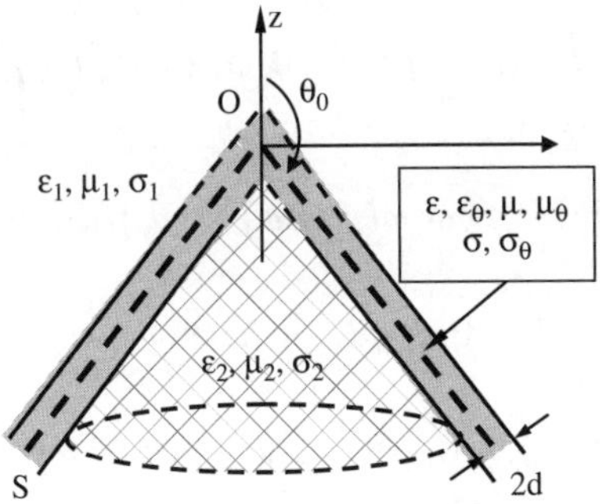

Figure 4.9. Geometrical structure of a simple conical layer far away from the tip.

A. Conical Layer in a Region Far Away from the Tip

In a region far away from the tip the layer seems to be formed by parallel surfaces (see Fig. 4.9). In this case it is better to write the shape function w as follows:

$$w = r(\theta - \theta_0). \tag{4.65a}$$

Indeed, by this choice the inner and outer boundaries of the layer corresponds, respectively, to

$$S_{\text{inner}}: \quad r(\theta - \theta_0) = +d, \qquad S_{\text{outer}}: \quad r(\theta - \theta_0) = -d. \tag{4.65b}$$

To find an explicit expression of (4.10a), one has to consider the following consecutive changes of coordinates:

$$(x, y, z) \to (r, \theta, \phi) \to (r, w, \phi).$$

Therefore one writes

$$\begin{aligned} dxdydz &= \frac{\partial(x, y, z)}{\partial(r, \theta, \phi)} dr d\theta d\phi = \frac{\partial(x, y, z)}{\partial(r, \theta, \phi)} \frac{\partial(r, \theta, \phi)}{\partial(r, w, \phi)} drdwd\phi \\ &= r \sin\theta drdwd\phi \end{aligned}$$

which yields also

$$\begin{aligned} < \mathbf{J}, \varphi > &= \iiint \mathbf{J}\varphi d\vartheta \\ &= \iiint \mathbf{J}(r, w, \phi)\varphi(r, w, \phi) r \sin\left(\frac{w}{r} + \theta_0\right) drdwd\phi \\ &= \iint \left\{ \int_{-d}^{d} \mathbf{J}(r, w, \phi)\varphi(r, w, \phi) \sin\left(\frac{w}{r} + \theta_0\right) dw \right\} rdrd\phi \\ &\approx \iint \left\{ \int_{-d}^{d} \mathbf{J}(r, 0, \phi)\varphi(r, 0, \phi) \sin\theta_0 \, dw \right\} rdrd\phi \\ &= \iint \{2d\mathbf{J}(r, 0, \phi)\varphi(r, 0, \phi) \sin\theta_0\} rdrd\phi \\ &= \iiint \left\{ 2d\mathbf{J}(r, 0, \phi)\varphi(r, w, \phi) \sin\left(\frac{w}{r} + \theta_0\right) \delta(w) dw \right\} rdrd\phi \\ &= \iiint \{2d\mathbf{J}(r, 0, \phi)\varphi(r, w, \phi)\delta(w)\} d\vartheta \\ &= < 2d\mathbf{J}(r, 0, \phi)\delta(w), \varphi > \end{aligned}$$

or, finally

$$\mathbf{J}(r,w,\phi) = 2d\mathbf{J}(r,0,\phi)\delta(w)$$
$$= \mathbf{J}_0(r,\phi)\delta(w), \qquad (4.66a)$$

where

$$\mathbf{J}_0(r,\phi) = d[\mathbf{J}(r,d-0,\phi) + \mathbf{J}(r,d+0,\phi)]. \qquad (4.66b)$$

From (4.66a,b) one concludes that the formulas (4.11a)–(4.17d) are all valid for this case also, provided that the shape function $w = z$ is replaced by $w = r(\theta - \theta_0)$ (see (4.65a)). Since

$$\text{grad}w = (\theta - \theta_0)\text{grad}r + r\text{grad}\theta,$$

on S one has

$$\text{grad}w = \mathbf{e}_\theta, \qquad \lambda = 1, \qquad (4.66c)$$

which is identical to (4.6) and (4.61). Therefore all the relations and results (i.e., Theorems 4.1–4.3; critical frequencies for impedance-type conditions, etc.) derived in Section 4.3.1 for planar sheets are also valid for conical structures in regions far away from the tip.

B. Conical Layer Near the Tip

Near the tip, S can be assumed to model a conical layer of width $(2\Delta\theta)$ as shown in Fig. 4.10. In this case, one has

$$w = \theta - \theta_0, \qquad (4.67a)$$

which yields

$$\text{grad}w = \frac{1}{r}\mathbf{e}_\theta, \qquad \mathbf{n} = \mathbf{e}_\theta, \qquad \lambda = \frac{1}{\text{r}}. \qquad (4.67b)$$

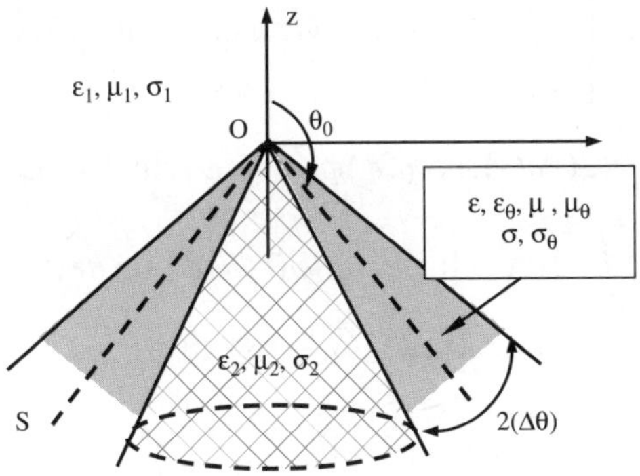

Figure 4.10. Geometrical structure of a simple conical layer near the tip.

Here (r, θ, ϕ) stand, as usual, the spherical polar coordinates. Since λ depends on the coordinates on S, the results obtained above for the case of $\lambda = 1$ (see Sections 4.3.1 and 4.3.2) will be changed substantially. Indeed, for a monochromatic wave of frequency ω, (4.5a–d) are read now:

$$\frac{1}{r}[[\mathbf{e}_\theta \times \mathbf{H}]] = \{\mathbf{J}_{0t} - i\omega\mathbf{P}_{0t}\} - \frac{1}{ri\omega\mu_0}\mathbf{e}_\theta \times \text{grad}\{ri\omega M_{0\theta})\} \quad (4.68a)$$

$$\frac{1}{r}[[\mathbf{e}_\theta \times \mathbf{E}]] = i\omega\mathbf{M}_{0t} + \frac{1}{ri\omega\varepsilon_0}\mathbf{e}_\theta \times \text{grad}\{r(J_{0\theta} - i\omega P_{0\theta})\} \quad (4.68b)$$

$$\frac{1}{r}[[\mathbf{e}_\theta \cdot \mathbf{D}]] = \rho_0 - \text{div}(\rho_1 r\mathbf{e}_\theta) - \text{div}\mathbf{P}_{0t} \quad (4.68c)$$

$$\frac{1}{r}[[\mathbf{e}_\theta \cdot \mathbf{B}]] = -\text{div}\mathbf{M}_{0t}. \quad (4.68d)$$

Let the constitutive parameters of the material filling the layer be $(\varepsilon, \varepsilon_\theta, \mu, \mu_\theta, \sigma, \sigma_\theta)$. Then, by repeating the analysis made in Section 4.3.1, one writes in the layer, for example,

$$\begin{aligned}
< \mathbf{J}, \varphi > &= \iiint \mathbf{J}\varphi d\vartheta \\
&= \iiint \mathbf{J}(r,\theta,\phi)\varphi(r,\theta,\phi)r^2 \sin\theta dr d\theta d\phi \\
&= \iint \left\{ \int_{\theta_0-\Delta\theta}^{\theta_0+\Delta\theta} \mathbf{J}(r,\theta,\phi)\varphi(r,\theta,\phi) \sin\theta \mathrm{d}\theta \right\} r^2 dr d\phi \\
&\approx \iint \left\{ \int_{\theta_0-\Delta\theta}^{\theta_0+\Delta\theta} \mathbf{J}(r,\theta_0,\phi)\varphi(r,\theta_0,\phi) \sin\theta_0 d\theta \right\} r^2 dr d\phi \\
&= \iint \{2(\Delta\theta)\mathbf{J}(r,\theta_0,\phi)\varphi(r,\theta_0,\phi) \sin\theta_0\} r^2 dr d\phi \\
&= \iiint \{2(\Delta\theta)\mathbf{J}(r,\theta_0,\phi)\varphi(r,\theta,\phi) \sin\theta\delta(\theta-\theta_0)\} r^2 dr d\phi d\theta \\
&= < 2(\Delta\theta)\mathbf{J}(r,\theta_0,\phi)\delta(\theta-\theta_0), \varphi(r,\theta,\phi) >,
\end{aligned}$$

which, with the assumption

$$\mathbf{J}(r,\theta_0,\phi) = \frac{1}{2}[\mathbf{J}(r,\theta_0 + \Delta\theta - 0,\phi) + \mathbf{J}(r,\theta_0 - \Delta\theta + 0,\phi)],$$

yields

$$\mathbf{J}(r,\theta,\phi) = \mathbf{J}_0(r,\phi)\delta(\theta - \theta_0). \tag{4.69a}$$

Here we put

$$\mathbf{J}_0(r,\phi) = (\Delta\theta)[\mathbf{J}(r,\theta_0 + \Delta\theta - 0,\phi) + \mathbf{J}(r,\theta_0 - \Delta\theta + 0,\phi)] \tag{4.69b}$$

(cf. (4.11a,b) and (4.64a,b)). This shows that (4.11c)–(4.15b) are valid for this case also with the convention $J_{0n} = J_{0\theta}$, and so on. It is worthwhile to note that the layer width $(2\Delta\theta)$ is now of dimension *radian* while in the cases considered in Sections 4.3.1, 4.3.2, and 4.3.3.A it was of dimension *meter*. This fact affects also the dimensions of $e, h, e_{1,2n}$, and $h_{1,2n}$.

If the material of the conical layer is not homogeneous such that the constitutive parameters depend on θ, then (4.14a)–(4.15b) are replaced by (4.9a–f). Assume this case and insert the expressions (4.9a–f) into (4.68a,b) to obtain

$$\mathbf{e}_\theta \times \mathbf{H}_t^{(1)} - \mathbf{e}_\theta \times \mathbf{H}_t^{(2)} + e_1 r\mathbf{E}_t^{(1)} + e_2 r\mathbf{E}_t^{(2)} + h_{1\theta}\mathbf{e}_\theta \times \operatorname{grad}(rH_\theta^{(1)}) + h_{2\theta}\mathbf{e}_\theta \times \operatorname{grad}(rH_\theta^{(2)}) = 0 \tag{4.70a}$$

$$h_1 r\mathbf{e}_\theta \times \mathbf{H}_t^{(1)} + h_2 r\mathbf{e}_\theta \times \mathbf{H}_t^{(2)} + \mathbf{E}_t^{(1)} - \mathbf{E}_t^{(2)} + e_{1\theta}\operatorname{grad}(rE_\theta^{(1)}) + e_{2\theta}\operatorname{grad}(rE_\theta^{(2)}) = 0. \tag{4.70b}$$

Note that (4.70b) consists of the cross-product of (4.68b) with $\mathbf{e}_\theta$. The relations (4.70a,b) constitute the couple of boundary conditions to be considered in solving the boundary-value problems connected with the above-mentioned material cone. As previously indicated also, in practical problems the boundary conditions are not always given in their original form (4.70a,b) but rather in certain linear combinations of them, namely,

$$(a_1 + \alpha_1 r)\mathbf{e}_\theta \times \mathbf{H}_t^{(1)} + (a_2 + \alpha_2 r)\mathbf{e}_\theta \times \mathbf{H}_t^{(2)} + (a_3 + \alpha_3 r)\mathbf{E}_t^{(1)} + (a_4 + \alpha_4 r)\mathbf{E}_t^{(2)} + a_5\mathbf{e}_\theta \times \operatorname{grad}(rH_\theta^{(1)}) + a_6\mathbf{e}_\theta \times \operatorname{grad}(r\,\mathrm{H}_\theta^{(2)}) + a_7\operatorname{grad}(rE_\theta^{(1)}) + a_8\operatorname{grad}(rE_\theta^{(2)}) = 0, \tag{4.71a}$$

$$(b_1 + \beta_1 r)\mathbf{e}_\theta \times \mathbf{H}_t^{(1)} + (b_2 + \beta_2 r)\mathbf{e}_\theta \times \mathbf{H}_t^{(2)} + (b_3 + \beta_3 r)\mathbf{E}_t^{(1)} + (b_4 + \beta_4 r)\mathbf{E}_t^{(2)} + \mathrm{b}_5\mathbf{e}_\theta \times \operatorname{grad}(rH_\theta^{(1)}) + \mathrm{b}_6\mathbf{e}_\theta \times \operatorname{grad}(rH_\theta^{(2)}) + b_7\operatorname{grad}(rE_\theta^{(1)}) + b_8\operatorname{grad}(rE_\theta^{(2)}) = 0. \tag{4.71b}$$

From (4.71a,b) one concludes that the boundary relation table C given in (4.27), which involves two rows, is replaced now by a table of four rows, namely,

$$C = \begin{bmatrix} a_1 & a_2 & a_3 & a_4 & a_5 & a_6 & a_7 & a_8 \\ \alpha_1 & \alpha_2 & \alpha_3 & \alpha_4 & — & — & — & — \\ b_1 & b_2 & b_3 & b_4 & b_5 & b_6 & b_7 & b_8 \\ \beta_1 & \beta_2 & \beta_3 & \beta_4 & — & — & — & — \end{bmatrix}. \tag{4.72a}$$

It is obvious that the entries in C have to satisfy the following relations:

$$a_1 + a_2 = 0, \qquad a_3 + a_4 = 0, \tag{4.72b}$$
$$b_1 + b_2 = 0, \qquad b_3 + b_4 = 0. \tag{4.72c}$$

For the original form written in (4.70a,b), C is as follows (cf. (4.28)):

$$C = \begin{bmatrix} 1 & -1 & 0 & 0 & h_{1\theta} & h_{2\theta} & 0 & 0 \\ 0 & 0 & e_1 & e_2 & — & — & — & — \\ 0 & 0 & 1 & -1 & 0 & 0 & e_{1\theta} & e_{2\theta} \\ h_1 & h_2 & 0 & 0 & — & — & — & — \end{bmatrix}. \tag{4.72d}$$

a. Example. Resistive-Type Conditions Near the Tips of Conically Curved Sheets. As an example, consider the material mentioned in Section 4.3.1.A, namely (4.18a,b), which yield

$$e_1 = e_2 = e \neq 0, \qquad h_1 = h_2 = h = 0, \tag{4.73a}$$
$$e_{1\theta} = e_{2\theta} = 0, \qquad h_{1\theta} = h_{2\theta} = 0. \tag{4.73b}$$

In such a case, (4.70a,b) are reduced to

$$\mathbf{e}_\theta \times [\mathbf{H}_t{}^{(1)} - \mathbf{H}_t{}^{(2)}] + er[\mathbf{E}_t^{(1)} + \mathbf{E}_t^{(2)}] = 0 \tag{4.73c}$$

and

$$\mathbf{E}_t^{(1)} - \mathbf{E}_t^{(2)} = 0, \tag{4.73d}$$

which can also be written as

$$\mathbf{E}_t^{(1)} = \mathbf{E}_t^{(2)}, \qquad \mathbf{e}_\theta \times [\mathbf{H}_t{}^{(1)} - \mathbf{H}_t{}^{(2)}] + \frac{1}{R} r\mathbf{E}_t^{(1)} = 0 \tag{4.73e}$$

with

$$R = \frac{1}{2e} = \frac{i}{2\omega(\Delta\theta)\{(\varepsilon - \varepsilon_0) + \frac{i\sigma}{\omega}\}}. \tag{4.73f}$$

The relations in (4.73e) are obviously the same as the relations in (4.19a), except for the factor (r) existing before $\mathbf{E}_t^{(1)}$. Since $\Delta\theta$ is of dimension *radian*, R taking place in (4.73f) is of dimension ohm × meter. Hence the equations in (4.73e) can be interpreted as the boundary conditions on a *conically curved resistive sheet*. The table C associated to this sheet is as follows:

$$C = \begin{bmatrix} 1 & -1 & 0 & 0 & 0 & 0 & 0 & 0 \\ 0 & 0 & 1/(2R) & 1/(2R) & - & - & - & - \\ 0 & 0 & 1 & -1 & 0 & 0 & 0 & 0 \\ 0 & 0 & 0 & 0 & - & - & - & - \end{bmatrix}. \tag{4.73g}$$

Now consider the dual of the resistive sheet (i.e., the so-called *magnetically conducting** sheet) defined through (4.21), which yields

$$e_1 = e_2 = e = 0, \qquad h_1 = h_2 = h \neq 0, \tag{4.74a}$$

$$e_{1\theta} = e_{2\theta} = 0, \qquad h_{1\theta} = h_{2\theta} = 0. \tag{4.74b}$$

In this case, (4.70a,b) are reduced to

$$\mathbf{e}_\theta \times \left[\mathbf{H}_t^{(1)} - \mathbf{H}_t^{(2)}\right] = 0 \tag{4.74c}$$

$$r\mathbf{e}_\theta \times \left[\mathbf{H}_t^{(1)} + \mathbf{H}_t^{(2)}\right] - 2G\left[\mathbf{E}_t^{(1)} - \mathbf{E}_t^{(2)}\right] = 0 \tag{4.74d}$$

with $G = -1/(2h)$. It is obvious that (4.74c,d) are identical to (4.22a,b) except that the unit vector $\mathbf{n}$ appearing in latter is replaced by $(\mathbf{e}_\theta/r)$. Therefore, the boundary conditions (4.74d) can be interpreted as the conditions on a *magnetically conducting conical sheet*. The table corresponding to this case is as follows:

$$C = \begin{bmatrix} 0 & 0 & 1 & -1 & 0 & 0 & 0 & 0 \\ -1/(2G) & -1/(2G) & 0 & 0 & - & - & - & - \\ 1 & -1 & 0 & 0 & 0 & 0 & 0 & 0 \\ 0 & 0 & 0 & 0 & - & - & - & - \end{bmatrix}. \tag{4.74e}$$

Remark. If the resistive boundary conditions are *defined* with (4.19a) on all sheets without considering their topological properties, then one arrives at an *unrealizable* constitutive parameters. Indeed, in such a case (4.19a) yields first (see also (4.71a,b))

$$a_1 + \alpha_1 r = 0, \quad a_2 + \alpha_2 r = 0, \quad a_3 + \alpha_3 r = -(a_4 + \alpha_4 r), \tag{4.75a}$$

*See the remark made in the footnote in page 78 for Section 4.3.1.A.

$$b_1 + \beta_1 r = -(b_2 + \beta_2 r), \quad b_3 + \beta_3 r + b_4 + \beta_4 r = \frac{1}{R}(b_1 + \beta_1 r), \tag{4.75b}$$

$$a_5 = a_6 = a_7 = a_8 = 0, \quad b_5 = b_6 = b_7 = b_8 = 0. \tag{4.75c}$$

Owing to the fact that $r \in [0, \infty)$ is a parameter and the conditions (4.72b,c) are to be satisfied, from (4.75a,b) one gets

$$a_1 = a_2 = 0, \qquad a_3 = -a_4 = \text{arbitrary}, \qquad a_5 = a_6 = a_7 = a_8 = 0,$$
$$\alpha_1 = \alpha_2 = 0, \qquad \alpha_3 = -\alpha_4 = \text{arbitrary},$$

and

$$b_1 = -b_2, \qquad b_1/R = b_3 + b_4 = 0, \qquad b_5 = b_6 = b_7 = b_8 = 0,$$
$$\beta_1 = -\beta_2, \qquad \beta_3 + \beta_4 = \beta_1/R.$$

If one assumes $R \neq \infty$, then from $b_1/R = 0$ one gets $b_1 = 0$ which requires a table of the form

$$C = \begin{bmatrix} 0 & 0 & a_3 & -a_3 & 0 & 0 & 0 & 0 \\ 0 & 0 & \alpha_3 & -\alpha_3 & - & - & - & - \\ 0 & 0 & b_3 & -b_3 & 0 & 0 & 0 & 0 \\ \beta_1 & -\beta_1 & \beta_3 & \beta_1/R - \beta_3 & - & - & - & - \end{bmatrix}, \tag{4.75d}$$

where $a_3, \alpha_3, b_3, \beta_1$, and β_3 are yet arbitrary constants while R is supposed to be given. If one tries to determine the constitutive parameters of the material filling the thin layer modeled by S, then from (4.79a) to be derived below one gets first of all the contradiction $0 = 1$.

Note also that the dimensions of R in (4.19a) and (4.73e) are not the same.

b. Impedance-Type Conditions Near the Tips of Conically Curved Sheets. Reconsider (4.70a,b) and try to eliminate $\mathbf{H}_t^{(1)}, \mathbf{E}_t^{(1)}, H_\theta^{(1)}$, and $E_\theta^{(1)}$ to obtain a relation involving only $\mathbf{H}_t^{(2)}, \mathbf{E}_t^{(2)}, H_\theta^{(2)}$, and $E_\theta^{(2)}$. To this end, multiply (4.70a) with $(h_1 r)$ and subtract from (4.70b) to obtain

$$\begin{aligned}(h_1 + h_2) r \mathbf{e}_\theta \times \mathbf{H}_t^{(2)} &- (1 + e_2 h_1 r^2)\mathbf{E}_t^{(2)} - h_1 h_{2\theta} r \mathbf{e}_\theta \times \operatorname{grad}(r H_\theta^{(2)}) \\ &+ e_{2\theta} \operatorname{grad}(r E_\theta^{(2)}) + (1 - e_1 h_1 r^2)\mathbf{E}_t^{(1)} + e_{1\theta} \operatorname{grad}(r E_\theta^{(1)}) \\ &- h_1 h_{1\theta} \mathbf{e}_\theta \times \operatorname{grad}(r H_\theta^{(1)}) = 0. \end{aligned} \tag{4.76}$$

It is obvious that the last two terms can be avoided with $e_{1\theta} = 0$ and $h_{1\theta} = 0$, whereas the component $\mathbf{E}_t^{(1)}$ can never disappear because of the existence of (r^2) in the coefficient associated with it. This shows that an *impedance-type condition is not compatible with the conical topology (near the tip) for materials considered in Section* 4.3.1.

Remark that the above-mentioned factor (r^2) is nothing but $(1/\lambda^2)$. Hence in the case of $\lambda = 1$, which is satisfied for planar or circularly cylindrical or spherical sheets as well as for the conical sheets far away from the tip, the coefficient of $\mathbf{E}_t^{(1)}$ becomes zero whenever the relation $e_1 h_1 = 1$ is satisfied (cf. (4.54a)). It is worthwhile to remark also that the dimensions of e and h are not the same for the conical sheet considered here and for planar (or circularly cylindrical or spherical or aforementioned conical) sheets.

c. Necessary and Sufficient Conditions for a Table *C* to Represent Boundary Conditions Near the Tip of a Conically Curved Material Sheet.

THEOREM 4.4 (Necessary Conditions)

Let C represent a couple of boundary relations on a certain simple conically curved material sheet. Then, in addition to (4.72b,c), one has

$$\text{Rank}\left\{\begin{bmatrix} a_3 & a_4 & \alpha_1 & \alpha_2 & a_7 & a_8 \\ b_3 & b_4 & \beta_1 & \beta_2 & b_7 & b_8 \end{bmatrix}\right\} \leq 1 \tag{4.77}$$

and

$$\text{Rank}\left\{\begin{bmatrix} a_1 & a_2 & \alpha_3 & \alpha_4 & a_5 & a_6 \\ b_1 & b_2 & \beta_3 & \beta_4 & b_5 & b_6 \end{bmatrix}\right\} \leq 1. \tag{4.78}$$

THEOREM 4.5 (Sufficient Conditions)

If the relations cited in Theorem 4.4 are fulfilled and $\Delta_3^1 = a_1 b_3 - a_3 b_1 \neq 0$, then C represents a uniquely determined simple conical material sheet.

Let us see the meaning and proof of these theorems in some detail. To this end consider first the Theorem 4.4. Under the assumptions of the theorem, (4.71a,b) consist of certain linear combinations of (4.70a,b). Hence, inversely, certain appropriate linear combinations of (4.71a,b) give (4.70a,b). Thus, with suitable constant coefficients A, B, A', and B', one has to write

$$A(a_1 + \alpha_1 r) + B(b_1 + \beta_1 r) = 1, \tag{4.79a}$$

$$A(a_2 + \alpha_2 r) + B(b_2 + \beta_2 r) = -1, \tag{4.79b}$$
$$Aa_7 + Bb_7 = 0, \tag{4.79c}$$
$$Aa_8 + Bb_8 = 0, \tag{4.79d}$$
$$A(a_3 + \alpha_3 r) + B(b_3 + \beta_3 r) = e_1 r, \tag{4.79e}$$
$$A(a_4 + \alpha_4 r) + B(b_4 + \beta_4 r) = e_2 r, \tag{4.79f}$$
$$Aa_5 + Bb_5 = h_{1n}, \tag{4.79g}$$
$$Aa_6 + Bb_6 = h_{2n} \tag{4.79h}$$

and

$$A'(a_3 + \alpha_3 r) + B'(b_3 + \beta_3 r)) = 1, \tag{4.80a}$$
$$A'(a_4 + \alpha_4 r) + B'(b_4 + \beta_4 r) = -1, \tag{4.80b}$$
$$A'a_5 + B'b_5 = 0, \tag{4.80c}$$
$$A'a_6 + B'b_6 = 0, \tag{4.80d}$$
$$A'(a_1 + \alpha_1 r) + B'(b_1 + \beta_1 r) = h_1 r, \tag{4.80e}$$
$$A'(a_2 + \alpha_2 r) + B'(b_2 + \beta_2 r) = h_2 r, \tag{4.80f}$$
$$A'a_7 + B'b_7 = e_{1n}, \tag{4.80g}$$
$$A'a_8 + B'b_8 = e_{2n}. \tag{4.80h}$$

Since $r \in [0, \infty)$ is an arbitrary parameter in (4.79a,b) and (4.79e,f), these latter are also equivalent to

$$Aa_1 + Bb_1 = 1, \tag{4.81a}$$
$$Aa_2 + Bb_2 = -1, \tag{4.81b}$$
$$Aa_3 + Bb_3 = 0, \tag{4.81c}$$
$$Aa_4 + Bb_4 = 0, \tag{4.81d}$$
$$A\alpha_1 + B\beta_1 = 0, \tag{4.81e}$$
$$A\alpha_2 + B\beta_2 = 0, \tag{4.81f}$$
$$A\alpha_3 + B\beta_3 = e_1, \tag{4.81g}$$
$$A\alpha_4 + B\beta_4 = e_2. \tag{4.81h}$$

Note that the equation in (4.81b) can also be replaced by

$$A(a_1 + a_2) + B(b_1 + b_2) = 0, \tag{4.81i}$$

which is obtained by adding (4.81a) and (4.81b) side by side and consists of an identity because of the relations (4.72b,c).

The constants (A, B) are to be determined through the 10 equations (4.79c,d), (4.81a), and (4.81c–f). After having determined these constants, the constitutive parameters e_1, e_2, h_{1n}, and h_{2n} are determined then through (4.81g,h) and (7.79g,h). In order for at least one of the constants A and B to be different from zero, the coefficients in the homogeneous equations (4.79c,d), and (4.81c–f) must satisfy the relation (4.77), with (4.72b,c) being taken into account. The numerical values of A and B are determined through one of the above-mentioned homogeneous equations and (4.81a).

As to the system (4.80a–h), the results associated with them can be obtained from the above results through the following substitutions of indices and quantities:

$$1 \leftrightarrow 3, \qquad 2 \leftrightarrow 4, \qquad 7 \leftrightarrow 5, \qquad 8 \leftrightarrow 6. \tag{4.82a}$$

and

$$A \to A', \quad B \to B', \quad e_1 \to h_1, \tag{4.82b}$$

$$e_2 \to h_2, \quad h_{1n} \to e_{1n}, \quad h_{2n} \to e_{2n}. \tag{4.82c}$$

This obviously gives the relations (4.78) and completes the proof of Theorem 4.4.

Now consider Theorem 4.5. By virtue of the assumption $\Delta_3^1 \neq 0$, from the equations (4.81a,c) one gets

$$A = \frac{b_3}{\Delta_3^1}, \qquad B = -\frac{a_3}{\Delta_3^1}, \tag{4.83}$$

which yield also (cf (4.81g,h) and (4.79g,h))

$$e_1 = \frac{b_3\alpha_3 - a_3\beta_3}{\Delta_3^1}, \qquad e_2 = \frac{b_3\alpha_4 - a_3\beta_4}{\Delta_3^1}, \tag{4.84a}$$

$$h_{1n} = \frac{b_3 a_5 - a_3 b_5}{\Delta_3^1}, \qquad h_{1n} = \frac{b_3 a_6 - a_3 b_6}{\Delta_3^1}. \tag{4.84b}$$

Similarly, by making the substitutions (4.82a–c) in (4.83)–(4.84b), one gets

$$A' = \frac{b_1}{\Delta_1^3}, \qquad B' = -\frac{a_1}{\Delta_1^3}, \tag{4.85}$$

$$h_1 = \frac{b_1\alpha_1 - a_1\beta_1}{\Delta_1^3}, \qquad h_2 = \frac{b_1\alpha_2 - a_1\beta_2}{\Delta_1^3}, \tag{4.86a}$$

$$e_{1n} = \frac{b_1 a_7 - a_1 b_7}{\Delta_1^3}, \qquad e_{1n} = \frac{b_1 a_8 - a_1 b_8}{\Delta_1^3}. \tag{4.86b}$$

d. Examples.

1. For the *interface* between simple media, one has

$$\mathbf{E}_t^{(1)} - \mathbf{E}_t^{(2)} = 0, \qquad \mathbf{e}_\theta \times [\mathbf{H}_t{}^{(1)} - \mathbf{H}_t{}^{(2)}] = 0.$$

Therefore the table C is as follows:

$$C = \begin{bmatrix} 0 & 0 & 1 & -1 & 0 & 0 & 0 & 0 \\ 0 & 0 & 0 & 0 & - & - & - & - \\ 1 & -1 & 0 & 0 & 0 & 0 & 0 & 0 \\ 0 & 0 & 0 & 0 & - & - & - & - \end{bmatrix}. \tag{4.87}$$

It satisfies all the requirements mentioned in Theorem 4.4. Since $\Delta_3^1 = -1 \neq 0$, the constitutive parameters can be uniquely determined through Theorem 4.5, which gives $e_1 = e_2 = h_1 = h_2 = e_{1\theta} = e_{1\theta} = h_{1\theta} = h_{1\theta} = 0$, as expected.

2. In the case of a *resistive sheet*, one has

$$\mathbf{E}_t^{(1)} - \mathbf{E}_t^{(2)} = 0, \qquad \mathbf{e}_\theta \times [\mathbf{H}_t{}^{(1)} - \mathbf{H}_t{}^{(2)}] + \frac{1}{R} r \mathbf{E}_t^{(1)} = 0$$

and

$$C = \begin{bmatrix} 1 & -1 & 0 & 0 & 0 & 0 & 0 & 0 \\ 0 & 0 & 1/(2R) & 1/(2R) & - & - & - & - \\ 0 & 0 & 1 & -1 & 0 & 0 & 0 & 0 \\ 0 & 0 & 0 & 0 & - & - & - & - \end{bmatrix}, \tag{4.88}$$

which yields $\Delta_3^1 = 1 \neq 0$ and

$$A = 1, \quad B = 0, \quad h_{1\theta} = h_{2\theta} = 0, \quad e_1 = e_2 = 1/(2\mathrm{R})$$
$$A' = 0, \quad B' = 1, \quad e_{1\theta} = e_{2\theta} = 0, \quad h_1 = h_2 = 0$$

as expected.

3. For the dual of the resistive sheet (i.e., the magnetically conducting sheet!) one writes

$$\mathbf{e}_\theta \times [\mathbf{H}_t{}^{(1)} - \mathbf{H}_t{}^{(2)}] = 0$$

and

$$r\mathbf{e}_\theta \times [\mathbf{H}_t{}^{(1)} + \mathbf{H}_t{}^{(2)}] - 2G[\mathbf{E}_\mathrm{t}^{(1)} - \mathbf{E}_\mathrm{t}^{(2)}] = 0,$$

which yields

$$C = \begin{bmatrix} 0 & 0 & 1 & -1 & 0 & 0 & 0 & 0 \\ -1/(2G) & -1/(2G) & 0 & 0 & — & — & — & — \\ 1 & -1 & 0 & 0 & 0 & 0 & 0 & 0 \\ 0 & 0 & 0 & 0 & — & — & — & — \end{bmatrix} \tag{4.89}$$

and

$$\Delta_3^1 = -1 \neq 0, \quad A = 0, \quad B = 1, \quad e_1 = 0, \quad e_2 = 0,$$
$$A' = 1, \quad B' = 0, \quad h_1 = -1/(2G), \quad h_2 = -1/(2G),$$
$$e_{1\theta} = 0, \qquad e_{2\theta} = 0, \qquad h_{1\theta} = 0, \qquad h_{2\theta} = 0.$$

4. The conditions on a *perfectly conducting cone* are written, as usual, as follows:

$$r(\mathbf{E}_t^{(1)} + \mathbf{E}_t^{(2)}) = 0, \qquad \mathbf{E}_t^{(1)} - \mathbf{E}_t^{(2)} = 0$$

The table corresponding to this case is

$$C = \begin{bmatrix} 0 & 0 & 0 & 0 & 0 & 0 & 0 & 0 \\ 0 & 0 & 1 & 1 & — & — & — & — \\ 0 & 0 & 1 & -1 & 0 & 0 & 0 & 0 \\ 0 & 0 & 0 & 0 & — & — & — & — \end{bmatrix}. \tag{4.90}$$

This epitomizes the cases that are not covered by Theorem 4.5 because $\Delta_3^1 = 0$. Note that this table is obtained from that of the resistive sheet by multiplying the first two rows of (4.88) by $(2R)$ and letting $R \to 0$.

PROBLEMS

1. For each of the following couple of relations, compose the table C and check if it satisfies the necessary conditions stated by (4.72b,c), (4.77), and (4.78):

(a) $\mathbf{E}_t^{(1)} - \mathbf{E}_t^{(2)} = 0, \qquad \mathbf{E}_t^{(1)} = 0,$

(b) $\mathbf{E}_t^{(1)} - \mathbf{E}_t^{(2)} = 0, \qquad r\mathbf{E}_t^{(1)} = 0,$

(c) $r\left(\mathbf{E}_t^{(1)} - \mathbf{E}_t^{(2)}\right) = 0, \qquad \mathbf{E}_t^{(1)} = 0,$

(d) $r(\mathbf{E}_t^{(1)} - \mathbf{E}_t^{(2)}) = 0, \qquad r\mathbf{E}_t^{(1)} = 0.$

2. Solve problem 1 for the following relations:

(a) $\mathbf{E}_t^{(1)} - \mathbf{E}_t^{(2)} = 0, \qquad \mathbf{e}_\theta \times \left[\mathbf{H}_t^{(1)} - \mathbf{H}_t^{(2)}\right] = 0,$

(b) $\mathbf{E}_t^{(1)} - \mathbf{E}_t^{(2)} = 0, \qquad r\mathbf{e}_\theta \times \left[\mathbf{H}_t^{(1)} - \mathbf{H}_t^{(2)}\right] = 0,$

(c) $r(\mathbf{E}_t^{(1)} - \mathbf{E}_t^{(2)}) = 0, \qquad \mathbf{e}_\theta \times \left[\mathbf{H}_t^{(1)} - \mathbf{H}_t^{(2)}\right] = 0,$

(d) $r\left(\mathbf{E}_t^{(1)} - \mathbf{E}_t^{(2)}\right) = 0, \qquad r\mathbf{e}_\theta \times \left[\mathbf{H}_t^{(1)} - \mathbf{H}_t^{(2)}\right] = 0.$

3. Solve problem 1 for the following relations:

(a) $\mathbf{E}_t^{(1)} - \mathbf{E}_t^{(2)} = 0, \qquad \mathbf{e}_\theta \times \left[\mathbf{H}_t^{(1)} - \mathbf{H}_t^{(2)}\right] + \frac{1}{R}\mathbf{E}_t^{(1)} = 0,$

(b) $\mathbf{E}_t^{(1)} - \mathbf{E}_t^{(2)} = 0, \qquad \mathbf{e}_\theta \times \left[\mathbf{H}_t^{(1)} - \mathbf{H}_t^{(2)}\right] + \frac{1}{R}r\mathbf{E}_t^{(1)} = 0,$

(c) $r\left(\mathbf{E}_t^{(1)} - \mathbf{E}_t^{(2)}\right) = 0, \qquad \mathbf{e}_\theta \times \left[\mathbf{H}_t^{(1)} - \mathbf{H}_t^{(2)}\right] + \frac{1}{R}\mathbf{E}_t^{(1)} = 0,$

(d) $r\left(\mathbf{E}_t^{(1)} - \mathbf{E}_t^{(2)}\right) = 0, \qquad \mathbf{e}_\theta \times \left[\mathbf{H}_t^{(1)} - \mathbf{H}_t^{(2)}\right] + \frac{1}{R}r\mathbf{E}_t^{(1)} = 0.$

4. Solve problem 1 for the following relations:

(a) $\mathbf{e}_\theta \times \left[\mathbf{H}_t^{(1)} - \mathbf{H}_t^{(2)}\right] = 0, \qquad \mathbf{e}_\theta \times \mathbf{H}_t^{(1)} - G\left[\mathbf{E}_t^{(1)} - \mathbf{E}_t^{(2)}\right] = 0,$

(b) $\mathbf{e}_\theta \times \left[\mathbf{H}_t^{(1)} - \mathbf{H}_t^{(2)}\right] = 0, \qquad r\mathbf{e}_\theta \times \mathbf{H}_t^{(1)} - G\left[\mathbf{E}_t^{(1)} - \mathbf{E}_t^{(2)}\right] = 0,$

(c) $r\mathbf{e}_\theta \times \left[\mathbf{H}_t^{(1)} - \mathbf{H}_t^{(2)}\right] = 0, \qquad \mathbf{e}_\theta \times \mathbf{H}_t^{(1)} - G\left[\mathbf{E}_t^{(1)} - \mathbf{E}_t^{(2)}\right] = 0,$

(d) $r\mathbf{e}_\theta \times \left[\mathbf{H}_t^{(1)} - \mathbf{H}_t^{(2)}\right] = 0, \qquad r\mathbf{e}_\theta \times \mathbf{H}_t^{(1)} - G\left[\mathbf{E}_t^{(1)} - \mathbf{E}_t^{(2)}\right] = 0.$

5. Solve problem 1 for the following relations:

(a) $\mathbf{E}_t^{(1)} - \mathbf{E}_t^{(2)} + A \text{ grad } r\left[\mathrm{E}_\theta^{(1)} + \mathrm{E}_\theta^{(2)}\right] = 0,$

$\mathbf{e}_\theta \times \left[\mathbf{H}_t^{(1)} - \mathbf{H}_t^{(2)}\right] + \frac{1}{2R}\left[\mathbf{E}_t^{(1)} + \mathbf{E}_t^{(2)}\right] = 0,$

(b) $\mathbf{E}_t^{(1)} - \mathbf{E}_t^{(2)} + A \text{ grad } r\left[\mathrm{E}_\theta^{(1)} + \mathrm{E}_\theta^{(2)}\right] = 0,$

$\mathbf{e}_\theta \times \left[\mathbf{H}_t^{(1)} - \mathbf{H}_t^{(2)}\right] + \frac{1}{2R}\mathrm{r}\left[\mathbf{E}_t^{(1)} + \mathbf{E}_t^{(2)}\right] = 0,$

(c) $\mathbf{E}_t^{(1)} - \mathbf{E}_t^{(2)} + A \text{ grad } r\left[\mathrm{E}_\theta^{(1)} + \mathrm{E}_\theta^{(2)}\right] = 0,$

$r\mathbf{e}_\theta \times \left[\mathbf{H}_t^{(1)} - \mathbf{H}_t^{(2)}\right] + \frac{1}{2R}\left[\mathbf{E}_t^{(1)} + \mathbf{E}_t^{(2)}\right] = 0,$

(d) $\mathbf{E}_{\mathrm{t}}^{(1)} - \mathbf{E}_{\mathrm{t}}^{(2)} + A \text{ grad } r \left[\mathrm{E}_{\theta}^{(1)} + \mathrm{E}_{\theta}^{(2)}\right] = 0,$

$r\mathbf{e}_{\theta} \times \left[\mathbf{H}_t^{(1)} - \mathbf{H}_t^{(2)}\right] + \frac{1}{2R}\mathrm{r}\left[\mathbf{E}_t^{(1)} + \mathbf{E}_t^{(2)}\right] = 0.$

6. Solve problem 1 for the following relations:

(a) $r\left[\mathbf{E}_{\mathrm{t}}^{(1)} - \mathbf{E}_{\mathrm{t}}^{(2)}\right] + A \text{ grad } r\left[E_{\theta}^{(1)} + E_{\theta}^{(2)}\right] = 0,$

$\mathbf{e}_{\theta} \times \left[\mathbf{H}_t^{(1)} - \mathbf{H}_t^{(2)}\right] + \frac{1}{2R}\left[\mathbf{E}_t^{(1)} + \mathbf{E}_t^{(2)}\right] = 0,$

(b) $r\left[\mathbf{E}_t^{(1)} - \mathbf{E}_t^{(2)}\right] + A \text{ grad } r\left[E_{\theta}^{(1)} + E_{\theta}^{(2)}\right] = 0,$

$\mathbf{e}_{\theta} \times \left[\mathbf{H}_t^{(1)} - \mathbf{H}_t^{(2)}\right] + \frac{1}{2R}r[\mathbf{E}_t^{(1)} + \mathbf{E}_t^{(2)}] = 0,$

(c) $r\left[\mathbf{E}_t^{(1)} - \mathbf{E}_t^{(2)}\right] + A \text{ grad } r\left[E_{\theta}^{(1)} + E_{\theta}^{(2)}\right] = 0,$

$r\mathbf{e}_{\theta} \times \left[\mathbf{H}_t^{(1)} - \mathbf{H}_t^{(2)}\right] + \frac{1}{2R}\left[\mathbf{E}_t^{(1)} + \mathbf{E}_t^{(2)}\right] = 0,$

(d) $r\left[\mathbf{E}_t^{(1)} - \mathbf{E}_t^{(2)}\right] + A \text{ grad } r\left[E_{\theta}^{(1)} + E_{\theta}^{(2)}\right] = 0,$

$r\mathbf{e}_{\theta} \times \left[\mathbf{H}_t^{(1)} - \mathbf{H}_t^{(2)}\right] + \frac{1}{2R}\mathrm{r}\left[\mathbf{E}_t^{(1)} + \mathbf{E}_t^{(2)}\right] = 0.$

7. Consider the relations given in problem 2(a) and multiply them with $(a_3 + \alpha_3 r)$ to get

$$(a_3 + \alpha_3 r)\mathbf{E}_t^{(1)} - (a_3 + \alpha_3 r)\mathbf{E}_t^{(2)} = 0,$$

and

$$(a_3 + \alpha_3 r)\mathbf{e}_{\theta} \times \mathbf{H}_t^{(1)} - (a_3 + \alpha_3 r)\mathbf{e}_{\theta} \times \mathbf{H}_t^{(2)} = 0.$$

Compose the table C pertinent to these conditions. Does it satisfy the necessary conditions stated by (4.72b,c), (4.77), and (4.78)?

CHAPTER 5

Discontinuities on a Moving Sheet

Consider now the case when the boundary S is in motion with respect to the observer. In such a case the boundary conditions as well as the compatibility relations are given by (3.12a–e) and (3.13a–e). But in application of these exact and general formulas there are some difficulties. Two main difficulties are as follows:

i. The total field is not, in general, time-harmonic even when the incident field is so.

ii. The shape of S is not known, in general, for an observer to which it is in motion.

The difficulty resulting from i is that it does not permit ones to use the advantages provided by the complex representation peculiar to time-harmonic fields. For example, it is not possible, in general, to extend and discuss the concept of impedance-type conditions.

As to the difficulty coming from ii, very frequently S consists of the boundary of a rigid body (for example, a sphere). An observer, say A, who moves with it knows its shape. But observer B, with respect to him it is in motion with velocity $\mathbf{v}(x, y, z, t)$, sees it quite differently. This shows that the shape function $w(x, y, z, t)$ mentioned in Ch.-3(3.5) is in fact not known beforehand. However, if the motion is

Discontinuties in the Electromagnetic Field, First Edition. By M. Mithat Idemen.

Published 2011 by John Wiley & Sons, Inc.

uniform, then the shape function $w(x, y, z, t)$ can be found very easily and exactly by using the known Lorentz transformation formulas of the Special Theory of Relativity. Hence, in what follows we will first consider this particular case separately. In this case it will also be interesting to derive the boundary conditions as well as the compatibility relations starting from the relations pertinent to boundaries at rest.

5.1 SPECIAL THEORY OF RELATIVITY

As is well known, the Special Theory of Relativity, which was discovered at the beginning of the last century, seemed to be born as a result of the long discussions on the problem of whether the *absolute motion* of the Earth can be detected through observations made on the Earth itself. In the earlier papers by Lorentz [32], Poincaré [33], and Einstein [34], published between 1904 and 1906, as well as in *almost* all papers and books published thereafter, the subject was presented in this way and based on the assumption that postulates the constancy of the velocity of the light regardless the situation of its source. To this end, Einstein first tried to redefine the notions of simultaneity, time, synchronous clocks, time interval, the length of a rod in a system at rest, the length in a moving system, and so on, and then stated two postulates that became thereafter the basis of the theory, namely:

i. The laws of physics are the same in all inertial (Galilean) reference frames.

ii. The speed of the light in empty space is the same in all inertial frames.

Even today the presentation of the subject is made, *in general*, in this way by dwelling on these postulates. The so-called Lorentz transformation, which constitutes the main basis of the theory, is derived by considering *only* the second of these postulates. This transformation, together with the first postulate, is applied then to the Electromagnetic Theory to reveal the rules that interrelate the expressions of the electromagnetic field in different Galilean systems, and to the classical Newtonian mechanics to *reestablish* it. In practice they are used to obtain the explicit expressions of solutions to complicated problems, connected with moving bodies, starting from their corresponding expressions supposed to be known in appropriate systems.

The above-mentioned approach, which was constantly repeated by Einstein in all his subsequent publications and adopted by *almost* all other writers, causes some confusing questions and, hence, makes it too hard to comprehend the theory and its issues such as the interrelation between the space coordinates and time, mass-energy equivalence, and so on. Among such questions we can note, for example:

a. Are the above-mentioned two postulates independent from each other? Is it possible to omit the second one to establish the theory.

b. Can anybody *reject* the Lorentz formulas *without rejecting* the Maxwell equations?

c. Why the Maxwell equations are correct *only* in Galilean systems?

It is also important to observe that sometimes laymen may think that there would be no problem (or the problem would not be detected) if one had not attempted to measure the velocity of the light meticulously. Although Einstein and his followers never gave up the light postulate in establishing the Theory of Special Relativity, its role had been subject to some objections since even earlier days of the theory. For example, in 1910 Ignatovski [35] had showed that the light postulate may be replaced by some kinematical assumptions to obtain a one-parameter family of space–time transformation groups under which space–time is invariant. The parameter is of the dimension of velocity and for *any* finite value of the parameter, the group is isomorphic to the Lorentz group. Furthermore, the inertial-frame-dependent formalism of Einstein was also objected by some scientists. For example, in 1914 Robb [36] tried to establish the theory in inertial-frame-independent form. With certain suitable postulates he obtained the geometry of Minkowski space in a purely geometrical manner. In Idemen [37] it was shown that only the first postulate of Einstein suffices to establish the Special Theory of Relativity if one assumes that the Maxwell equations are correct in all Galilean reference systems. Thus, from one side the Special Relativity becomes a natural issue of the Maxwell equations, and from the other side the questions and confusions mentioned above become removed. To this end, a rather simple problem connected with the Maxwell equations was formulated and solved in three different coordinate systems, and the results were compared. The problem consisted of finding the electromagnetic field created by a point charge while the coordinate systems in question were as follows:

a. A system in which the Maxwell equations are supposed to be valid (a Galilean system) while the charged point is in motion with constant (vector) velocity.

b. A system attached to the charged particle.

c. A system that is not Galilean.

In what follows we will present the solutions pertinent to the *first two cases* very briefly in order.

5.1.1 The Field Created by a Uniformly Moving Point Charge

Consider a Galilean system of reference, say a Cartesian coordinate system $Oxyz$, in which a point charge of amount Q makes a uniform motion in the direction Ox with a constant velocity equal to v. If one assumes that at the time $t = 0$ the charge is at the origin, then the corresponding charge and current densities become

$$\rho = Q\delta(x - vt)\delta(y)\delta(z) \tag{5.1}$$

and

$$\mathbf{J} = Qv\delta(x - vt)\delta(y)\delta(z)\mathbf{e}_x, \tag{5.2}$$

respectively. Since one *assumes* that the Maxwell equations

$$\text{curl}\mathbf{H} - \varepsilon_0\frac{\partial}{\partial t}\mathbf{E} = \mathbf{J}, \qquad \text{curl}\mathbf{E} + \mu_0\frac{\partial}{\partial t}\mathbf{H} = 0, \tag{5.3a,b}$$

$$\varepsilon_0 \ \text{div}\mathbf{E} = \rho, \qquad \text{div}\mathbf{H} = 0 \tag{5.3c,d}$$

are valid (in the *vacuum*) in such a reference system, the electromagnetic field can be expressed through a couple of scalar and vector potentials, say $V(x, y, z, t)$ and $\mathbf{A}(x, y, z, t)$, as follows*:

$$\mathbf{E} = -\text{grad}V - \partial\mathbf{A}/\partial t, \qquad \mathbf{H} = (1/\mu_0)\text{curl}\mathbf{A}. \tag{5.3e}$$

*See Section 3.1.1, problem 2.

From (5.3b) and (5.3d) it is obvious that the couple of potentials mentioned here is not unique. In order to avoid the ambiguity, we assume also that the so-called Lorentz gauge condition

$$\mathrm{div}\mathbf{A} + \varepsilon_0\mu_0\frac{\partial}{\partial t}V = 0 \tag{5.3f}$$

is met by the potentials we will consider. Thus the explicit expressions of the potentials created by the source given by (5.1)–(5.2) can be obtained rather easily by computing the retarded potential integrals, namely,*

$$V(x,y,z,t) = \frac{Q}{4\pi\varepsilon_0}\iiint \delta\left(\xi - v\left[\mathrm{t} - \frac{R}{c_0}\right]\right)\delta(\eta)\delta(\zeta)\frac{d\xi d\eta d\varsigma}{R} \tag{5.4}$$

and

$$A_1(x,y,z,t) = \frac{v}{c_0^2}V(x,y,z,t), \qquad A_2 = 0, \qquad A_3 = 0. \tag{5.5}$$

Here R denotes the distance between the observation point (x, y, z) and the volume element at (ξ,η,ζ), namely,

$$R = \{(x-\xi)^2 + (y-\eta)^2 + (z-\zeta)^2\}^{1/2}. \tag{5.6}$$

In (5.5) the quantities with sub-indices 1, 2, and 3 signify the Cartesian components of a vector, namely, $\mathbf{A} = (A_1, A_2, A_3)$. By definition of the Dirac distribution, in (5.4) the integrations with respect to η and ζ are immediate and yield

$$V(x,y,z,t) = \frac{Q}{4\pi\varepsilon_0}\int_{-\infty}^{\infty}\delta\left(\xi - vt + \frac{v}{c_0}\sqrt{(x-\xi)^2+y^2+z^2}\right)\frac{d\xi}{\sqrt{(x-\xi)^2+y^2+z^2}}. \tag{5.7}$$

To compute this integral, we make the substitution (cf. Jones [3, Section 3.1])

$$\xi - vt + \frac{v}{c_0}\sqrt{(x-\xi)^2+y^2+z^2} = \lambda, \tag{5.8a}$$

*See Section 3.1.1, problem 3.

which yields

$$\frac{d\lambda}{d\xi} = 1 - \frac{v}{c_0}\frac{x-\xi}{\sqrt{(x-\xi)^2+y^2+z^2}} > 0. \tag{5.8b}$$

This shows that $\lambda = \lambda(\xi)$ is a monotonic increasing function of ξ. Hence to $\xi = -\infty$ and $\xi = +\infty$ there correspond $\lambda = -\infty$ and $\lambda = +\infty$, respectively, which shows that at a unique point $\xi = \xi_1$ one can have $\lambda = 0$. Thus one gets

$$\begin{aligned} V(x,y,z,t) &= \frac{Q}{4\pi\varepsilon_0}\int_{-\infty}^{\infty}\delta(\lambda)\frac{1}{\sqrt{(x-\xi)^2+y^2+z^2}}\frac{d\lambda}{1-\frac{v}{c_0}\frac{x-\xi}{\sqrt{(x-\xi)^2+y^2+z^2}}} \\ &= \frac{Q}{4\pi\varepsilon_0}\frac{1}{\sqrt{(x-\xi_1)^2+y^2+z^2}-\frac{v}{c_0}(x-\xi_1)}. \end{aligned} \tag{5.9a}$$

An explicit expression of ξ_1 is obtained as follows from (5.8a) by inserting $\lambda = 0$:

$$\xi_1 = \frac{1}{1-c_0^2/v^2}\left[-\left(\frac{c_0^2 t}{v} - x\right) + \frac{c_0}{v}\sqrt{(x-vt)^2+(1-v^2/c_0^2)(y^2+z^2)}\right],$$

which reduces (5.9a) into

$$\begin{aligned} V(x,y,z,t) &= \frac{Q}{4\pi\varepsilon_0}\frac{1}{\frac{v}{c_0}(\xi_1 - x)+\frac{c_0}{v}(vt-\xi_1)} \\ &= \frac{Q}{4\pi\varepsilon_0}\frac{1}{\sqrt{(x-vt)^2+(1-v^2/c_0^2)(y^2+z^2)}}. \end{aligned} \tag{5.9b}$$

This expression of the scalar potential, together with (5.5) and (5.3e), provides us an explicit expression of the electromagnetic field created by the point charge making the uniform motion mentioned above.

5.1.2 The Expressions of the Field in a Reference System Attached to the Charged Particle

Consider now a reference system $O'x'y'z'$ whose origin O' moves with the above-mentioned point charge such that the axes $O'x'$, $O'y'$, and $O'z'$ are parallel to Ox, Oy, and Oz, respectively. Since this system is also Galilean (*it is known that this is in accordance with the composition rule of velocities to be derived from the Lorentz transformation to be obtained later on*), by assumption the Maxwell equations are also valid there with the same constants ε_0, μ_0, and c_0. Thus, if we denote all the quantities observed in the system $O'x'y'z'$ with the same letter used above but with an upper sign ($'$), then we write

$$\rho'(x', y', z', t') = Q\delta(x')\delta(y')\delta(z') \tag{5.10}$$

and

$$\mathbf{J}'(x', y', z', t') = 0, \tag{5.11}$$

which yield

$$V'(x', y', z', t') = \frac{Q}{4\pi\varepsilon_0}\frac{1}{\sqrt{(x'^2 + y'^2 + z'^2)}} \tag{5.12a}$$

and

$$\mathbf{A}'(x', y', z', t') \equiv 0. \tag{5.12b}$$

These expressions of the potential functions, together with (5.3e), give the explicit expression of the field created by the same charge in the system $O'x'y'z'$:

$$\mathbf{E}' = -\mathrm{grad}V', \qquad \mathbf{H}' = 0. \tag{5.13}$$

5.1.3 Lorentz Transformation Formulas [37]

Now compare (5.3e), (5.5), and (5.9b) with (5.12a)–(5.13). They consist of the expressions of the same phenomenon in the reference systems (x, y, z, t) and (x', y', z', t'). If one assumes that they can be transformed into one another by a *universal* transformation rule, then one writes for the electric field, for example,

$$\mathbf{E} = \mathcal{L}\mathcal{M}_e\{\mathbf{E}'(x', y', z', t'),\ \mathbf{H}'(x', y', z', t')\}. \tag{5.14a}$$

Here $\mathcal{M}_e$ refers to the transformation rule pertinent to the field components (i.e., $(\mathbf{E}', \mathbf{H}') \rightarrow \mathbf{E}$) while $\mathcal{L}$ denotes the transformation of the coordinates (i.e., $(x', y', z', t') \rightarrow (x, y, z, t)$). The *universal* rule $\mathcal{M}_e$ is inevitably independent of the position of the observation point. Therefore it has no effect on $(x', y', z', t') \rightarrow (x, y, z, t)$. If one *assumes* also that

$$y' = y, \qquad z' = z, \tag{5.14b}$$

which means that the motion of the reference system $O'x'y'z'$ with respect to $Oxyz$ consists of a translation *without rotation and expansion*, then we write

$$y'^2 + z'^2 = y^2 + z^2 \equiv \zeta. \tag{5.14c}$$

From (5.9b) and (5.12a) it is obvious that both sides of (5.14a) depend on the same $\zeta \geq 0$. It is also obvious that both sides of (5.14a) can be analytically continued into the complex ζ-plane. Hence, in accordance with the *principle of permanence of functional equations* [38], these continuations satisfy the continuation of (5.14a). In other words, (5.14a) is valid not only for $\zeta \geq 0$ but also for all complex ζ. Therefore, the transformation rule $\mathcal{L}$ has to transform the branch point $\varsigma = -x'^2$ appearing on the right-hand side to the branch point $\varsigma = -(x - vt)^2/(1 - v^2/c_0^2)$ which exists on the left-hand side. Thus one gets

$$x' = \frac{x - vt}{\sqrt{1 - v^2/c_0^2}}, \qquad y' = y, \qquad z' = z. \tag{5.15a}$$

These relations determine $\mathcal{L}$ except the expression of t'.

According to Einstein [34], the transformation rule $\mathcal{L}$ is not peculiar only to the electromagnetic phenomena to use as a receipt to transform the field components from one Galilean system to another, but rather general and associated with the reference systems $Oxyz$ (including t) and $O'x'y'z'$ (including t'). Therefore (5.15a) is also valid to express (x, y, z) in terms of (x', y', z', t'). If one *assumes* that the former seems to be in motion with velocity $(-\mathbf{v})$ with respect to the latter,* then one writes also

$$x = \frac{x' + vt'}{\sqrt{1 - v^2/c_0^2}}, \qquad y = y', \qquad z = z'. \tag{5.15b}$$

*This assumption is in accordance with the rule of composition of velocities, which is based on (5.15d).

From (15a,b) one gets

$$t = \frac{t' + (v/c_0^2)x'}{\sqrt{1 - v^2/c_0^2}}, \qquad t' = \frac{t - (v/c_0^2)x}{\sqrt{1 - v^2/c_0^2}}. \tag{5.15c}$$

Equations (5.15a–c) can be summarized in a matrix equation as follows:

$$\begin{bmatrix} x \\ y \\ z \\ c_0 t \end{bmatrix} = \begin{bmatrix} 1/\sqrt{1 - v^2/c_0^2} & 0 & 0 & (v/c_0)/\sqrt{1 - v^2/c_0^2} \\ 0 & 1 & 0 & 0 \\ 0 & 0 & 1 & 0 \\ (v/c_0)/\sqrt{1 - v^2/c_0^2} & 0 & 0 & 1/\sqrt{1 - v^2/c_0^2} \end{bmatrix} \begin{bmatrix} x' \\ y' \\ z' \\ c_0 t' \end{bmatrix}. \tag{5.15d}$$

Equation (5.15d) is nothing but the above-mentioned transformation $\mathcal{L}$. It is referred to as the Lorentz transformation. It claims that from physical phenomena point of view the space is a four-dimensional variety (Minkowski space).

PROBLEMS

1. Let S' be the plane with the shape function $w' \equiv \cos\alpha x' + \sin\alpha \cos\beta y' + \sin\alpha \sin\beta z' - d$, where the geometrical interpretations of the angles α and β are shown in the accompanying figure.

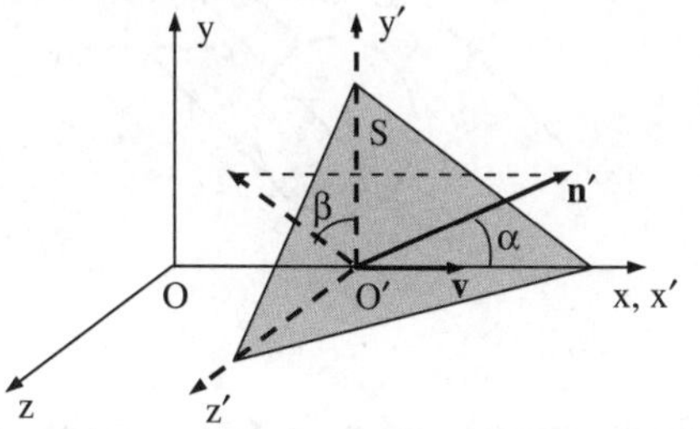

(a) Find the expression of the shape function seen from the reference system $Oxyz$.

(b) Show that the unit normal vector **n** to S is

$$\lambda \mathbf{n} = \frac{\cos\alpha}{\sqrt{1-(v/c_0)^2}}\mathbf{e}_x + \sin\alpha\cos\beta\mathbf{e}_y + \sin\alpha\sin\beta\mathbf{e}_z$$

with

$$\lambda = \sqrt{\frac{\cos^2\alpha}{1-(v/c_0)^2} + \sin^2\alpha}.$$

2. Let S' be a portion (or whole) of the a circular cylinder which moves in a direction normal to itself (see the accompanying figure).

(a) Write the shape function in the system $O'x'y'z'$.

(b) What is the expression of the shape function in the system $Oxyz$? Show that S seems to an observer lying in the system $Oxyz$ as an elliptic cylinder. What are the diameters of the elliptic cylinder?

(c) Show that the unit normal vector **n** to S is given by

$$\lambda \mathbf{n} = 2\frac{x - vt}{1-(v/c_0)^2}\mathbf{e}_x + 2y\mathbf{e}_y$$

where one has

$$\lambda = \frac{2}{\sqrt{1-(v/c_0)^2}}\sqrt{R^2 - \frac{v^2}{c_0^2}y^2}.$$

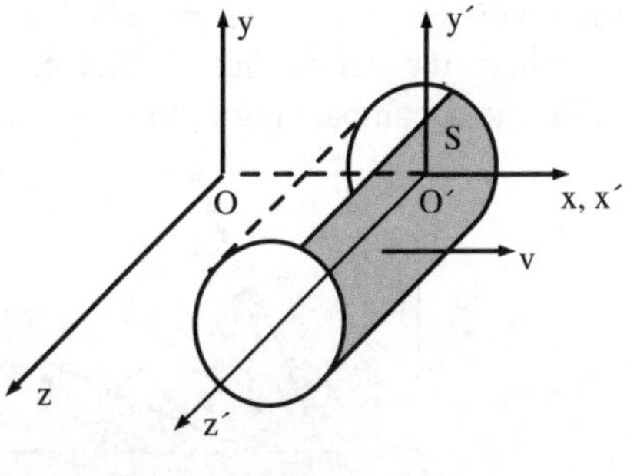

5.1.4 Transformation of the Electromagnetic Field

By using (5.5), (5.9b), and (5.12a,b), we can easily convince ourselves that the vector of components $(A_1, A_2, A_3, V/c_0)$ is also a vector in the above-mentioned four-dimensional space:

$$\begin{bmatrix} A_1 \\ A_2 \\ A_3 \\ V/c_0 \end{bmatrix} = \begin{bmatrix} 1/\sqrt{1-v^2/c_0^2} & 0 & 0 & (v/c_0)/\sqrt{1-v^2/c_0^2} \\ 0 & 1 & 0 & 0 \\ 0 & 0 & 1 & 0 \\ (v/c_0)/\sqrt{1-v^2/c_0^2} & 0 & 0 & 1/\sqrt{1-v^2/c_0^2} \end{bmatrix} \begin{bmatrix} A'_1 \\ A'_2 \\ A'_3 \\ V'/c_0 \end{bmatrix}. \tag{5.16}$$

If one computes the field components in the system $Oxyz$ through (5.3) and transform the resulting expressions into $O'x'y'z'$ by using (5.15d) and (5.16), then one gets the following formulas (see Idemen [37]):

$$E_1 = E'_1, \qquad E_2 = \frac{1}{\sqrt{1-v^2/c_0^2}}[E'_2 + vB'_3] \tag{5.17a}$$

$$E_3 = \frac{1}{\sqrt{1-v^2/c_0^2}}[E'_3 - vB'_2], \tag{5.17a$'$}$$

$$B_1 = B'_1, \qquad B_2 = \frac{1}{\sqrt{1-v^2/c_0^2}}\left[B'_2 - \frac{v}{c_0^2}E'_3\right], \tag{5.17b}$$

$$B_3 = \frac{1}{\sqrt{1-v^2/c_0^2}}\left[B'_3 + \frac{v}{c_0^2}E'_2\right], \tag{5.17b$'$}$$

$$D_1 = D'_1, \qquad D_2 = \frac{1}{\sqrt{1-v^2/c_0^2}}\left[D'_2 + \frac{v}{c_0^2}H'_3\right], \tag{5.17c}$$

$$D_3 = \frac{1}{\sqrt{1-v^2/c_0^2}}\left[D'_3 - \frac{v}{c_0^2}H'_2\right], \tag{5.17c$'$}$$

$$H_1 = H'_1, \qquad H_2 = \frac{1}{\sqrt{1-v^2/c_0^2}}[H'_2 - vD'_3], \tag{5.17d}$$

$$H_3 = \frac{1}{\sqrt{1-v^2/c_0^2}}[H'_3 + vD'_2], \tag{5.17d$'$}$$

$$J_1 = \frac{1}{\sqrt{1 - v^2/c_0^2}}[J'_1 + \rho' v], \qquad J_2 = J'_2, \tag{5.17e}$$

$$J_3 = J'_3, \qquad \rho = \frac{1}{\sqrt{1 - v^2/c_0^2}}\left[\rho' + \frac{v}{c_0^2}J'_1\right]. \tag{5.17e'}$$

PROBLEM

1. Show that the formulas (5.17a–e′) can also be written as follows:

$$\mathbf{E}' = \frac{1}{\sqrt{1 - v^2/c_0^2}}\mathbf{E} + \left[1 - \frac{1}{\sqrt{1 - v^2/c_0^2}}\right](\mathbf{E}\cdot\mathbf{v})\frac{1}{v^2}\mathbf{v} + \frac{1}{\sqrt{1 - v^2/c_0^2}}\mathbf{v}\times\mathbf{B},$$

$$\mathbf{D}' = \frac{1}{\sqrt{1 - v^2/c_0^2}}\mathbf{D} + \left[1 - \frac{1}{\sqrt{1 - v^2/c_0^2}}\right](\mathbf{D}\cdot\mathbf{v})\frac{1}{v^2}\mathbf{v} + \frac{1/c_0^2}{\sqrt{1 - v^2/c_0^2}}\mathbf{v}\times\mathbf{H},$$

$$\mathbf{H}' = \frac{1}{\sqrt{1 - v^2/c_0^2}}\mathbf{H} + \left[1 - \frac{1}{\sqrt{1 - v^2/c_0^2}}\right](\mathbf{H}\cdot\mathbf{v})\frac{1}{v^2}\mathbf{v} - \frac{1}{\sqrt{1 - v^2/c_0^2}}\mathbf{v}\times\mathbf{D},$$

$$\mathbf{B}' = \frac{1}{\sqrt{1 - v^2/c_0^2}}\mathbf{B} + \left[1 - \frac{1}{\sqrt{1 - v^2/c_0^2}}\right](\mathbf{B}\cdot\mathbf{v})\frac{1}{v^2}\mathbf{v} - \frac{1/c_0^2}{\sqrt{1 - v^2/c_0^2}}\mathbf{v}\times\mathbf{E},$$

$$\mathbf{J}' = \mathbf{J} - \left[1 - \frac{1}{\sqrt{1 - v^2/c_0^2}}\right](\mathbf{J}\cdot\mathbf{v})\frac{1}{v^2}\mathbf{v} - \frac{1}{\sqrt{1 - v^2/c_0^2}}\rho\mathbf{v},$$

$$\rho' = \frac{1}{\sqrt{1 - v^2/c_0^2}}\left[\rho - \frac{1}{c_0^2}(\mathbf{J}\cdot\mathbf{v})\right].$$

5.2 DISCONTINUITIES ON A UNIFORMLY MOVING SURFACE [75]

Consider an electromagnetic field created by certain sources in the vacuum which involves various bodies of different constitutive parameters. Let one of these bodies be Ω'. Consider also a reference

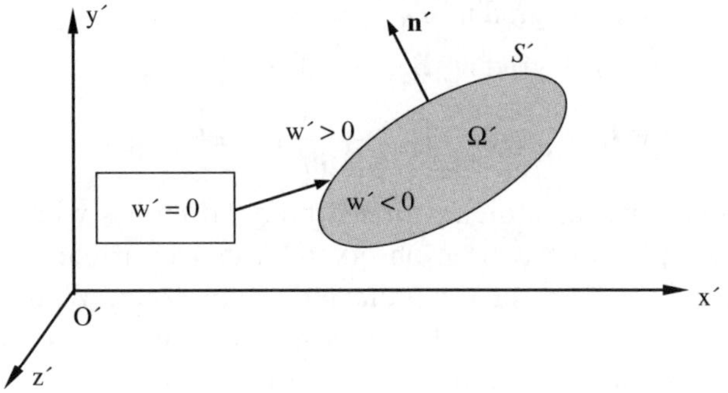

Figure 5.1. A body Ω' and its boundary S' in vacuum.

system $O'x'y'z'$ attached to Ω' such that the boundary of Ω', say S', is defined through a given continuously differentiable function $w'(x', y', z')$ by $w'(x', y', z') = 0$ (see Fig. 5.1). Without loss of generality, we can assume that inside and outside Ω' one has $w' < 0$ and $w' > 0$, respectively. As we will see later on, the function $w'(x', y', z')$ will play basic roles in the relations we want to derive. It is referred to as the *shape function* of S'. In what follows we will denote all quantities observed in $O'x'y'z'$ by primes ($'$). Thus the general formulas (3.12a–e) and (3.13a–e) are reduced to (in what follows we replace $\lambda\mathbf{n}$ by grad w'):

$k = 0 \Rightarrow$

$$[[\text{grad}'w' \times \mathbf{H}']] = -\text{curl}'\mathbf{H}'_0 + \frac{\partial}{\partial t'}\mathbf{D}'_0 + \mathbf{J}'_0 \tag{5.18a}$$

$$[[\text{grad}'w' \times \mathbf{E}']] = -\text{curl}'\mathbf{E}'_0 - \frac{\partial}{\partial t'}\mathbf{B}'_0 \tag{5.18b}$$

$$[[\text{grad}'w' \cdot \mathbf{D}']] = \rho'_0 - \text{div}'\mathbf{D}'_0 \tag{5.18c}$$

$$[[\text{grad}'w' \cdot \mathbf{B}']] = -\text{div}'\mathbf{B}'_0 \tag{5.18d}$$

$$[[\text{grad}'w' \cdot \mathbf{J}']] = -\frac{\partial}{\partial t'}\rho'_0 - \text{div}'\mathbf{J}'_0 \tag{5.18e}$$

$k \geq 1 \Rightarrow$

$$\text{curl}'\mathbf{H}'_k + \text{grad}'w' \times \mathbf{H}'_{k-1} - \frac{\partial}{\partial t'}\mathbf{D}'_k = \mathbf{J}'_k \tag{5.19a}$$

$$\text{curl}'\mathbf{E}'_k + \text{grad}'w' \times \mathbf{E}'_{k-1} + \frac{\partial}{\partial t'}\mathbf{B}'_k = 0 \tag{5.19b}$$

$$\mathrm{div}'\mathbf{D}'_k + \mathrm{grad}'w' \cdot \mathbf{D}'_{k-1} = \rho'_k \tag{5.19c}$$

$$\mathrm{div}'\mathbf{B}'_k + \mathrm{grad}'w' \cdot \mathbf{B}'_{k-1} = 0 \tag{5.19d}$$

$$\mathrm{div}'\mathbf{J}'_k + \mathrm{grad}'w' \cdot \mathbf{J}'_{k-1} + \frac{\partial}{\partial t'}\rho'_k = 0. \tag{5.19e}$$

Now assume that the aforementioned body Ω' moves with a *constant* velocity $\mathbf{v}$ with respect to an observer located in a reference system represented by a Cartesian coordinate system $Oxyz$. Without loss of generality, we can suppose that $\mathbf{v}$ as well as the x' axis are parallel to the x axis and at the time $t = 0$ we have $t = t' = 0$ and $O \equiv O'$ (see Fig. 5.2). In this case the body seems to the observer staying in $Oxyz$ quite differently. If we denote the surface of the body, seen by an observer in $Oxyz$, by S, then S is determined through the equation $w(x, y, z, t) = 0$ which consists of the transformation of $w'(x', y', z') = 0$ by the Lorentz formulas (5.15d), namely,

$$w(x, y, z, t) = w'(\gamma(x - vt), y, z) = 0. \tag{5.20a}$$

Here γ stands for the following square root

$$\gamma = \frac{1}{\sqrt{1 - (v/c_0)^2}}. \tag{5.20b}$$

It is important to observe that from (5.15a) one gets

$$\frac{\partial w}{\partial t} = -\gamma \mathrm{v} \frac{\partial w'}{\partial x'}, \qquad x = \text{constant}, \tag{5.21a}$$

and

$$\frac{\partial w}{\partial x} = \gamma \frac{\partial w'}{\partial x'}, \qquad t = \text{constant}, \tag{5.21b}$$

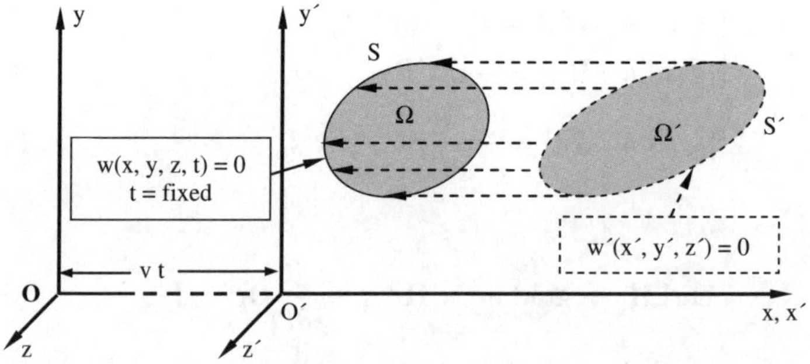

Figure 5.2. A body seen from the reference systems $Oxyz$ and $O'x'y'z'$.

which yield also

$$\frac{\partial w}{\partial t} + \mathbf{v} \cdot \mathrm{grad} w = \frac{\partial w}{\partial t} + \mathrm{v}\frac{\partial w}{\partial x} = 0. \tag{5.21c}$$

To find the expressions of the boundary conditions satisfied on S, one has to replace the field components and source densities $\mathbf{E}'$, $\mathbf{J}'$, and so on, which take place in (5.18a)–(5.19e) by their expressions in terms of quantities measured by an observer in the system $Oxyz$ (see (5.17a–e′)). Furthermore, the derivatives $(\partial/\partial x')$, $(\partial/\partial y')$, $(\partial/\partial z')$, and $(\partial/\partial t')$, which take place in (5.18a)–(5.19e), have also to be computed in terms of $(\partial/\partial x)$, $(\partial/\partial y)$, $(\partial/\partial z)$, and $(\partial/\partial t)$. With simple and straightforward computations, one obtains

$$\frac{\partial}{\partial x'} = \gamma\left[\frac{\partial}{\partial x} + \frac{v}{c_0^2}\frac{\partial}{\partial t}\right], \quad t' = \text{constant} \tag{5.22a}$$

$$\frac{\partial}{\partial t'} = \gamma\left[\frac{\partial}{\partial t} + v\frac{\partial}{\partial x}\right], \quad x' = \text{constant} \tag{5.22b}$$

$$\frac{\partial}{\partial y'} = \frac{\partial}{\partial y}, \qquad \frac{\partial}{\partial z'} = \frac{\partial}{\partial z}. \tag{5.22c}$$

If these rules are applied to the function $w'(x', y', z') \equiv w(x(x', t'), y, z, t(x', t'))$, then by virtue of (5.21b) one writes also

$$\begin{aligned}\frac{\partial w'}{\partial x'} &= \gamma\left[\frac{\partial w}{\partial x} + \frac{\mathrm{v}}{c_0^2}\frac{\partial w}{\partial t}\right]\\ &= \gamma\left[\frac{\partial w}{\partial x} - \mathrm{v}\frac{v}{c_0^2}\frac{\partial w}{\partial x}\right]\\ &= \frac{1}{\gamma}\frac{\partial w}{\partial x}, \qquad t' = \text{constant},\end{aligned} \tag{5.22d}$$

as expected (cf. (5.21a)).

5.2.1 Transformation of the Universal Boundary Conditions [75]

Consider first the equation (5.18a) and write its projections on the coordinate axes $O'x'$, $O'y'$, and $O'z'$. If we transform the field components and source densities by (5.17a–e′) while the derivatives are computed through (5.22a–d), after some obvious simplifications we obtain (see problem 1 below)

$$\left[\left[\frac{\partial w}{\partial y}H_z - \frac{\partial w}{\partial z}H_y\right]\right] - v\left[\left[\frac{\partial w}{\partial y}D_y + \frac{\partial w}{\partial z}D_z\right]\right] = \left\{\frac{\partial H_{0y}}{\partial z} - \frac{\partial H_{0z}}{\partial y}\right\} + v \operatorname{div} \mathbf{D}_0 + \frac{\partial D_{0x}}{\partial t} + \mathrm{J}_{0x} - v\rho, \tag{5.23a}$$

$$\left[\left[\frac{\partial w}{\partial z}H_x - \frac{\partial w}{\partial x}H_z\right]\right] + v\left[\left[\frac{\partial w}{\partial x}D_y\right]\right] = \left\{\frac{\partial H_{0z}}{\partial x} - \frac{\partial H_{0x}}{\partial z}\right\} + \frac{\partial D_{0y}}{\partial t} + \mathrm{J}_{0y}, \tag{5.23b}$$

$$\left[\left[\frac{\partial w}{\partial x}H_y - \frac{\partial w}{\partial y}H_x\right]\right] + v\left[\left[\frac{\partial w}{\partial x}D_z\right]\right] = \left\{\frac{\partial H_{0x}}{\partial y} - \frac{\partial H_{0y}}{\partial x}\right\} + \frac{\partial D_{0z}}{\partial t} + J_{0z}. \tag{5.23c}$$

It is obvious that (5.23b) and (5.23c) have a cyclic symmetry property that (5.13a) does not have. This is an expected result because the motion is assumed to be in the direction of Ox axis, and makes the equations (5.23a–c) reference-depending. However, by considering the fact that the fields taking place in (5.23a) are not arbitrary but rather solutions to the Maxwell equations, (5.23a) can also be replaced by an equation which has the cyclic symmetry in question. To find this equation, consider (5.18c), which is transformed to

$$-\frac{v}{c_0^2}\left[\left[\frac{\partial w}{\partial y}H_z - \frac{\partial w}{\partial z}H_y\right]\right] + \left[\left[\frac{\partial w}{\partial y}D_y + \frac{\partial w}{\partial z}D_z\right]\right] + \frac{1}{\gamma^2}\left[\left[\frac{\partial w}{\partial x}D_x\right]\right] = \rho_0 - \frac{v}{c_0^2}J_{0x} - \operatorname{div}\mathbf{D}_0 - \frac{v}{c_0^2}\frac{\partial D_{0x}}{\partial t} - \frac{v}{c_0^2}\left\{\frac{\partial H_{0y}}{\partial z} - \frac{\partial H_{0z}}{\partial y}\right\}. \tag{5.24}$$

If we multiply both sides of this equation by $\boldsymbol{v}$ and add to (5.23a), then we get

$$\left[\left[\frac{\partial w}{\partial y}H_z - \frac{\partial w}{\partial z}H_y\right]\right] + v\left[\left[\frac{\partial w}{\partial x}D_x\right]\right] = \left\{\frac{\partial H_{0y}}{\partial z} - \frac{\partial H_{0z}}{\partial y}\right\} + \frac{\partial D_{0x}}{\partial t} + J_{0x}, \tag{5.25}$$

which has the cyclic symmetry mentioned above. A close scrutiny shows that (5.23b), (5.23c), and (5.25) consist of the projections of the following vector equation on the Ox, Oy, and Oz axes, respectively:

$$[[\mathrm{grad}w \times \mathbf{H}]] + (\mathbf{v}\cdot\mathrm{grad}w)[[\mathbf{D}]] = -\mathrm{curl}\mathbf{H}_0 + \frac{\partial}{\partial t}\mathbf{D}_0 + \mathbf{J}_0. \quad (5.26a)$$

This is the equation that replaces (5.18a) in the system $Oxyz$.

Now multiply (5.23a) by (v/c_0^2) and add to (5.24), to get

$$[[\mathrm{grad}w\cdot\mathbf{D}]] = \rho_0 - \mathrm{div}\mathbf{D}_0, \quad (5.26b)$$

which replaces (5.18c) in the system $Oxyz$.

It is interesting and worthwhile to note that (5.26a) is *not the direct transformation* of (5.18a) because in order to arrive it we had also to use (5.18c). Similarly, (5.26b) is also *not the direct transformation* of (5.18c); to get it we had to use also (5.23a) (or (5.18a)).

Quite similarly, starting from (5.18b) and (5.18d), one obtains

$$[[\mathrm{grad}w \times \mathbf{E}]] - (\mathbf{v}\cdot\mathrm{grad}w)[[\mathbf{B}]] = -\mathrm{curl}\mathbf{E}_0 - \frac{\partial}{\partial t}\mathbf{B}_0 \quad (5.26c)$$

and

$$[[\mathrm{grad}w\cdot\mathbf{B}]] = -\mathrm{div}\mathbf{B}_0 \quad (5.26d)$$

while (5.18e) is transformed to

$$[[\mathrm{grad}w\cdot\mathbf{J}]] - (\mathbf{v}\cdot\mathrm{grad}w)[[\rho]] = -\frac{\partial}{\partial t}\rho_0 - \mathrm{div}\mathbf{J}_0. \quad (5.26e)$$

It is obvious that (5.26a–e) are the same as (3.12a–e), which were obtained directly from the Maxwell equations.

PROBLEMS

1. **(a)** Apply the transformation rules to (5.18a) and show (5.23a–c).

   (b) Transform (5.18c) and show (5.24).

   (c) Repeat the operations mentioned in (a) and (b) for (5.18b) and (5.18d) to get (5.26c,d).

2. Apply the transformation rules to (5.18e) and show (5.26e).

5.2.2 Transformation of the Compatibility Relations [75]

By repeating the procedure applied in the previous subsection, one gets from (5.19a–e) the following compatibility relations seen from the system $Oxyz$ (in what follows $k \geq 1$):

$$\text{curl}\mathbf{H}_k + \text{grad}w \times \mathbf{H}_{k-1} - \frac{\partial}{\partial t}\mathbf{D}_k + (\mathbf{v}\cdot\text{grad}w)\mathbf{D}_{k-1} = \mathbf{J}_k, \quad (5.27a)$$

$$\text{curl}\mathbf{E}_k + \text{grad}w \times \mathbf{E}_{k-1} + \frac{\partial}{\partial t}\mathbf{B}_k - (\mathbf{v}\cdot\text{grad}w)\mathbf{B}_{k-1} = 0, \quad (5.27b)$$

$$\text{div}\mathbf{D}_k + \text{grad}w\cdot\mathbf{D}_{k-1} = \rho_k, \quad (5.27c)$$

$$\text{div}\mathbf{B}_k + \text{grad}w\cdot\mathbf{B}_{k-1} = 0, \quad (5.27d)$$

$$\text{div}\mathbf{J}_k + \text{grad}w\cdot\mathbf{J}_{k-1} + \frac{\partial}{\partial t}\rho_k - (\mathbf{v}\cdot\text{grad}w)\rho_{k-1} = 0. \quad (5.27e)$$

The system (5.27a–e) is nothing but the compatibility equations (3.13a–e) which were obtained directly from the Maxwell equations. They are also independent of the coordinate system.

PROBLEMS

1. **(a)** Apply the transformation rules to (5.19a) and (5.19c) and show (5.27a) and (5.27c).

   (b) Repeat the operations mentioned in (a) for (5.19b) and (5.19d) to get (5.27b) and (5.27d).

2. Apply the transformation rules to (5.19e) and show (5.27e).

5.2.3 Some Simple Examples

The quantities of zero sub-indices appearing in (5.26a–e) consist of *values* concentrated on S (i.e., singular parts). They can also be expressed in terms of the limiting field values measured on both sides of S. If S consists of an *ordinary interface* between two media, they are determined rather easily by considering the compatibility relations (5.27a–e) together with the fact that there is no physical existence of S. If one of the regions separated by S is perfectly conducting, then ρ_0 and $\mathbf{J}_0$ become different from zero. Since they are not known beforehand, in this case the relations (5.26a) and (5.26b) cannot be used as boundary conditions to solve the boundary-value problem which aims to determine the explicit expression of the field. The boundary-value problem in question is solved by using the homogeneous relations (5.26c) and (5.26d). After having thus determined the field functions explicitly, ρ_0 and $\mathbf{J}_0$ can be found by using the relations (5.26a) and (5.26b). When S models a very thin layer, the quantities with zero sub-indices show the average of the corresponding quantities (see, for ex., Idemen [19]). To illustrate the issues of the results obtained above, in what follows we want to consider some particular cases that are also of importance from practical points of view.

A. Boundary Conditions on the Interface of Two Simple Media

Suppose that two simple media are in contact across a regular interface S which moves in accordance with (5.20a). In this case on the interface one observes an accumulation of charges (of density ρ_S) and a current (of density $\mathbf{J}_S$). That means that all the quantities concentrated on S but $\rho_0, \mathbf{J}_0, \mathbf{E}_0, \mathbf{H}_0, \mathbf{D}_0$, and $\mathbf{B}_0$ are naught, which reduce the compatibility equations (5.27a–e) into the following:

$$\mathbf{n} \times \mathbf{H}_0 + v_n \mathbf{D}_0 = 0 \tag{5.28a}$$

$$\mathbf{n} \times \mathbf{E}_0 - v_n \mathbf{B}_0 = 0 \tag{5.28b}$$

$$\mathbf{n} \cdot \mathbf{D}_0 = 0 \tag{5.28c}$$

$$\mathbf{n} \cdot \mathbf{B}_0 = 0 \tag{5.28d}$$

$$\mathbf{n} \cdot \mathbf{J}_0 - v_n \rho_0 = 0. \tag{5.28e}$$

Note that $\mathbf{n}$ may depend on (x, y, z) and t on S.

By considering the identities $\mathbf{P}_0 \equiv 0$ and $\mathbf{M}_0 \equiv 0$, one writes

$$\mathbf{D}_0 = \varepsilon_0 \mathbf{E}_0, \qquad \mathbf{B}_0 = \mu_0 \mathbf{H}_0, \tag{5.29}$$

which reduce (5.28c,d) to

$$E_{0n} = 0, \qquad H_{0n} = 0. \tag{5.30}$$

These latter equations are also compatible with (5.28a,b). If we use these results in (5.28a,b), then for the tangential components we write

$$\mathbf{n} \times \mathbf{H}_{0t} + v_n \varepsilon_0 \mathbf{E}_{0t} = 0, \tag{5.31a}$$

$$\mathbf{n} \times \mathbf{E}_{0t} - v_n \mu_0 \mathbf{H}_{0t} = 0. \tag{5.31b}$$

By cross-multiplication of (5.31b) with **n**, one obtains

$$\mathbf{E}_{0t} + v_n \mu_0 \mathbf{n} \times \mathbf{H}_{0t} = 0. \tag{5.32}$$

The compatibility condition of the homogeneous linear equations in (5.31a) and (5.32) requires

$$(v_n)^2 = (c_0)^2. \tag{5.33}$$

Since in accordance with the Special Theory of Relativity one always has $v_n < c_0$, (5.33) never meets and yields

$$\mathbf{E}_{0t} = 0, \qquad \mathbf{H}_{0t} = 0, \tag{5.34}$$

which, along with (5.30), yield

$$\mathbf{E}_0 = 0, \qquad \mathbf{H}_0 = 0. \tag{5.35}$$

Now let us insert (5.35) into the boundary conditions (5.26a–e) and use the new definitions

$$\rho_S = \frac{1}{\lambda}\rho_0, \qquad \mathbf{J}_S = \frac{1}{\lambda}\mathbf{J}_0. \tag{5.36}$$

The result becomes

$$[[\mathbf{n} \times \mathbf{H}_t]] + v_n[[\mathbf{D}_t]] = \mathbf{J}_{st}, \tag{5.37a}$$

$$v_n[[D_n]] = J_{sn}, \tag{5.37a$'$}$$

$$[[\mathbf{n} \times \mathbf{E}_t]] - v_n[[\mathbf{B}_t]] = 0, \tag{5.37b}$$

$$[[\mathbf{n} \cdot \mathbf{D}]] = \rho_s, \tag{5.37c}$$

$$[[\mathbf{n} \cdot \mathbf{B}]] = 0, \tag{5.37d}$$

and

$$\begin{aligned} [[\mathbf{n} \cdot \mathbf{J}]] - v_n[[\rho]] &= -\frac{1}{\lambda}\frac{\partial}{\partial t}\rho_0 - \frac{1}{\lambda}\mathrm{div}\mathbf{J}_0 \\ &\equiv -\frac{1}{\lambda}\frac{\partial}{\partial t}(\lambda\rho_S) - \frac{1}{\lambda}\mathrm{div}(\lambda\mathbf{J}_s). \end{aligned} \tag{5.37e}$$

Note that the boundary relations (5.37a–d) are equivalent to the relations given in Namias [39] (see also references 40–44).

From (5.36) and (5.28e) (or from (5.37a′) and (5.37c)), one gets also

$$J_{sn} \equiv \frac{1}{\lambda}\mathrm{J}_{0n} = \frac{1}{\lambda}v_n\rho_0 = \rho_s v_n, \tag{5.38}$$

which shows that the surface currents in the normal directions consist of the motion of the surface charges. At this stage we want to consider three cases separately.

a. The Case when Both Regions Separated by *S* are Lossless Dielectrics. If an observer moving with S sees both regions to be lossless (i.e., $\sigma_1 = \sigma_2 = 0$), then he observes also $\mathbf{J}'_S \equiv 0$ and $\rho'_S \equiv 0$, which yield

$$\mathbf{J}_S \equiv 0, \qquad \rho_S \equiv 0. \tag{5.39}$$

Thus (5.37a–e) are reduced to

$$[[\mathbf{n} \times \mathbf{H}_t]] + v_n[[\mathbf{D}_t]] = 0, \tag{5.40a}$$

$$[[\mathbf{n} \times \mathbf{E}_t]] - v_n[[\mathbf{B}_t]] = 0, \tag{5.40b}$$

$$[[\mathbf{n} \cdot \mathbf{D}]] = 0, \tag{5.40c}$$

$$[[\mathbf{n} \cdot \mathbf{B}]] = 0, \tag{5.40d}$$

and

$$[[\mathbf{n} \cdot \mathbf{J}]] - v_n[[\rho]] = 0. \tag{5.40e}$$

b. The Case when Both Regions Separated by *S* are Lossy Dielectrics. If the regions separated by S seems to a co-moving observer to be lossy dielectrics with finite conductivities σ_1 and σ_2, then one has $\mathbf{J}'_S \equiv 0$ while $\rho'_S \not\equiv 0$ which yields (cf. Section 5.1.4, problem 1)

$$\mathbf{J}_s = \frac{\rho'_S}{\sqrt{1 - v^2/c_0^2}}\mathbf{v}, \qquad \rho_s = \frac{\rho'_S}{\sqrt{1 - v^2/c_0^2}} \tag{5.41a}$$

or

$$\mathbf{J}_s = \rho_s \mathbf{v}. \tag{5.41b}$$

In this case, (5.37a–e) become

$$[[\mathbf{n} \times \mathbf{H}_t]] + v_n[[\mathbf{D}_t]] + [[D_n]]\mathbf{n} \times (\mathbf{n} \times \mathbf{v}) = 0, \tag{5.42a}$$

$$[[\mathbf{n} \times \mathbf{E}_t]] - v_n[[\mathbf{B}_t]] = 0, \tag{5.42b}$$

$$[[D_n]] = \rho_s, \tag{5.42c}$$

$$[[B_n]] = 0 \tag{5.42d}$$

and

$$[[J_n]] - v_n[[\rho]] = -\frac{1}{\lambda}\frac{\partial}{\partial t}(\lambda \rho_s) - \frac{1}{\lambda}\mathrm{div}(\lambda \rho_s \mathbf{v}). \tag{5.42e}$$

c. The Case when One of the Regions Separated by *S* is Perfectly Conducting. When the region (1) is filled with a perfectly conducting material, an observer moving with S observes

$$\mathbf{E}'^{(1)} = \mathbf{D}'^{(1)} = \mathbf{H}'^{(1)} = \mathbf{B}'^{(1)} = \mathbf{J}'^{(1)} = \rho'^{(1)} \equiv 0 \tag{5.43a}$$

while

$$\mathbf{J}'_S \not\equiv 0, \qquad \rho'_S \not\equiv 0. \tag{5.43b}$$

Thus (5.37a–e) are reduced to the following:

$$\mathbf{n} \times \mathbf{H}_t^{(2)} + v_n \mathbf{D}_\mathrm{t}^{(2)} = \mathbf{J}_{st}, \tag{5.44a}$$

$$\mathbf{n} \times \mathbf{E}_t^{(2)} - v_n \mathbf{B}_\mathrm{t}^{(2)} = 0, \tag{5.44b}$$

$$D_n^{(2)} = \rho_S, \tag{5.44c}$$

$$B_n^{(2)} = 0, \tag{5.44d}$$

and

$$J_n^{(2)} - v_n \rho^{(2)} = -\frac{1}{\lambda}\frac{\partial}{\partial t}(\lambda \rho_s) - \frac{1}{\lambda}\mathrm{div}(\lambda \mathbf{J}_s). \tag{5.44e}$$

From (5.44a) and cross-multiplication of (5.44b) by $\mathbf{v}$ one writes

$$\mathbf{n} \times \mathbf{H}_t^{(2)} + v_n \varepsilon_2 \mathbf{E}_t^{(2)} = \mathbf{J}_{st}, \qquad \mathbf{E}_t^{(2)} + v_n \mu_2 \mathbf{H}_t^{(2)} = 0, \tag{5.44f}$$

which yield

$$\left(1 - v_n^2/c_0^2\right) \mathbf{n} \times \mathbf{H}_t^{(2)} = \mathbf{J}_{st}. \tag{5.44g}$$

The homogeneous boundary relations (5.44b) and (5.44d) are used, in general, to solve the boundary-value problem, which attempts to determine the explicit expression of the field in question while the nonhomogeneous relations (5.44a) (or (5.44g)) and (5.44c) are used to reveal the surface charges and currents concentrated on S. From (5.44b) (from (5.44a) and (5.44g)) one concludes that *a perfectly conducting body seems as a body covered by an impedance sheet if it is in motion. The impedance in question depends not only on the value of the velocity but also on the time and shape of the body*, which affects $v_n = (\mathbf{v} \cdot \mathrm{grad} w)/\lambda$.

d. An Illustrative Application. Assume that a perfectly conducting planar sheet is moving in *vacuum* with a constant velocity v in a direction that is normal to itself and is illuminated by a linearly polarized plane wave whose electric field is parallel to the sheet. In this case we can choose the coordinate axes so that the shape function

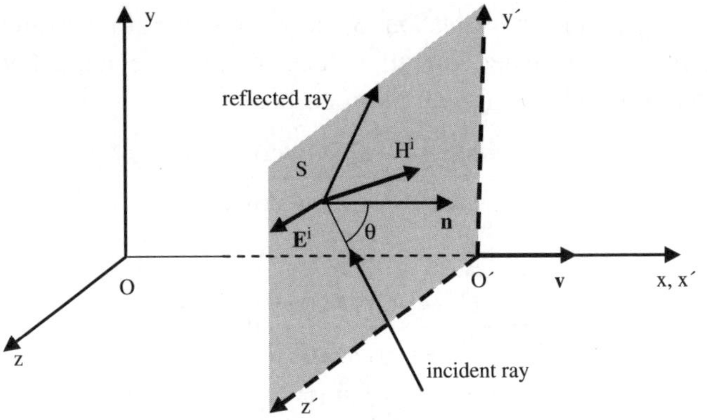

Figure 5.3. A uniformly moving plane.

in the reference system $Oxyz$ becomes $w = x - vt$ while the incident wave is (see Fig. 5.3)

$$\mathbf{E}^{\mathrm{i}} = \Phi(-x\cos\theta + y\sin\theta - c_0 t)\mathbf{e}_z, \tag{5.45a}$$

$$\mathbf{H}^{\mathrm{i}} = \sqrt{\frac{\varepsilon_0}{\mu_0}}\Phi(-x\cos\theta + y\sin\theta - c_0 t)[\sin\theta\mathbf{e}_x + \cos\theta\mathbf{e}_y], \tag{5.45b}$$

where $\Phi(\lambda)$ is a given function. In what follows we will assume $\theta \in (0, \pi/2)$. Hence when $v > 0$, the incident wave propagates toward the plane while in the case of $v < 0$ it follows the plane. To find the reflected wave, we have to consider first the homogeneous boundary conditions (5.44b) and (5.44d), which yield explicitly

$$E_y^{(2)} - v\mu_2 H_z^{(2)} = 0, \qquad E_z^{(2)} + v\mu_2 H_y^{(2)} = 0, \qquad H_x^{(2)} = 0. \tag{5.46}$$

Since the reflector plane as well as the incident wave are all planar, one can expect that the reflected wave also is planar and linearly polarized. Therefore we can try to find the reflected wave as follows:

$$\mathbf{E}^{r} = f(x\cos\tilde{\theta} + y\sin\tilde{\theta} - c_0 t)\mathbf{e}_z, \tag{5.47a}$$

$$\mathbf{H}^{r} = \sqrt{\frac{\varepsilon_0}{\mu_0}} f(x\cos\tilde{\theta} + y\sin\tilde{\theta} - c_0 t)[\sin\tilde{\theta}\mathbf{e}_x - \cos\tilde{\theta}\mathbf{e}_y]. \tag{5.47b}$$

The first condition in (5.46) is obviously an identity while the second one yields

$$\left[1 + \frac{v}{c_0}\cos\theta\right]\Phi(y\sin\theta - (c_0 + v\cos\theta)t) + \left[1 - \frac{v}{c_0}\cos\tilde{\theta}\right] f(y\sin\tilde{\theta} - (c_0 - v\cos\tilde{\theta})t) \equiv 0, \tag{5.48a}$$

with $\mu_2 = \mu_0$ and $x = vt$ on S being taken into account. Since (5.48a) is to be satisfied for all values of the independent variables $t \in (-\infty, \infty)$ and $y \in (-\infty, \infty)$, one has

$$\frac{\sin\theta}{\sin\tilde{\theta}} = \frac{c_0 + v\cos\theta}{c_0 - v\cos\tilde{\theta}} \tag{5.48b}$$

and

$$\begin{aligned} f(\eta) &= -\frac{1 + (v/c_0)\cos\theta}{1 - (v/c_0)\cos\tilde{\theta}}\Phi\left(\frac{\sin\theta}{\sin\tilde{\theta}}\eta\right) \\ &= -\frac{\sin\theta}{\sin\tilde{\theta}}\Phi\left(\frac{\sin\theta}{\sin\tilde{\theta}}\eta\right). \end{aligned} \tag{5.48c}$$

Thus, finally one gets the reflected wave

$$E_z^r = -\frac{\sin\theta}{\sin\tilde{\theta}}\Phi\left(\frac{\sin\theta}{\sin\tilde{\theta}}\{x\cos\tilde{\theta} + y\sin\tilde{\theta} - c_0 t\}\right) \tag{5.49a}$$

$$H_x^r = -\sqrt{\frac{\varepsilon_0}{\mu_0}}\sin\theta\Phi\left(\frac{\sin\theta}{\sin\tilde{\theta}}\{x\cos\tilde{\theta} + y\sin\tilde{\theta} - c_0 t\}\right) \tag{5.49b}$$

$$H_y^r = \sqrt{\frac{\varepsilon_0}{\mu_0}}\cos\tilde{\theta}\frac{\sin\theta}{\sin\tilde{\theta}}\Phi\left(\frac{\sin\theta}{\sin\tilde{\theta}}\{x\cos\tilde{\theta} + y\sin\tilde{\theta} - c_0 t\}\right). \tag{5.49c}$$

Note that (5.45b), (5.47b), and (5.48c) make the third boundary condition in (5.46) an identity.

The direction of propagation of the reflected wave (i.e., the angle $\tilde{\theta}$), is determined through (5.48b), which can also be written as

$$\sin\left(\frac{\theta - \tilde{\theta}}{2}\right) = \frac{v}{c_0}\sin\left(\frac{\theta + \tilde{\theta}}{2}\right)$$

or

$$\tan\frac{\tilde{\theta}}{2} = \frac{1 - v/c_0}{1 + v/c_0}\tan\frac{\theta}{2}. \tag{5.50a}$$

From (5.50a) it is obvious that $\tilde{\theta}$ is never equal to θ. One has $\tilde{\theta} < \theta$ if $v > 0$ while $\tilde{\theta} > \theta$ if $v < 0$. This is the well-known *aberration* effect. When $\theta \to 0$, one gets $\tilde{\theta} \to 0$ and

$$\frac{\sin\theta}{\sin\tilde{\theta}} \to \frac{\tan(\theta/2)}{\tan(\tilde{\theta}/2)} = \frac{1 + v/c_0}{1 - v/c_0}. \tag{5.50b}$$

If $\Phi(\eta)$ is periodic with period p, then both the incident and the reflected waves are periodic with periods T and T', respectively, such that

$$c_0 T = p, \qquad \frac{\sin\theta}{\sin\tilde{\theta}} c_0\, T' = p.$$

From these one gets

$$T' = \frac{\sin\tilde{\theta}}{\sin\theta} T, \qquad \omega' = \frac{\sin\theta}{\sin\tilde{\theta}} \omega. \tag{5.51a,b}$$

Here ω and ω' stand, respectively, for the angular frequencies of the incident and reflected waves. The difference between them is the known *Doppler effect*.* When $v > 0$—that is, when the plate moves toward the incident wave—from (5.50a,b) and (5.51b) one concludes that both the amplitude and the frequency of the reflected wave are larger than those of the incident wave. Since the density of the transmitted energy is proportional to the square of the amplitude, from this one concludes also that the energy associated with the reflected wave is larger than the energy of the incident wave. The difference is provided by the motion of the plate. In other words, to move the plate one has to spend a force that contributes to the energy of the reflected wave. Inversely, when $v < 0$—that is, when the plate follows the incident wave—both the amplitude and frequency of the reflected wave are smaller than those associated with the incident wave. That means that the energy density of the reflected wave is also smaller than the density of the incident wave. The difference in the energy contributes to the motion of the plate. From these explanations one concludes that the motion of a perfectly conducting plate can be performed more easily in the direction of propagation than in the inverse direction.

By using (5.45a, b) and (5.49a,b) in (5.44a) and (5.44c), one gets the explicit expressions of the charges and currents excited on plate S. They are as follows:

$$\rho_S = \varepsilon_0 (E_x^i + E_x^r)_{x=vt} \equiv 0 \tag{5.52}$$

*It is interesting to observe from (5.51b) and (5.49a) that the reflection coefficient is equal to the Doppler shift factor ($\sin\theta / \sin\tilde{\theta}$).

and

$$
\begin{aligned}
J_{Sz} &= [(H_y^i + H_y^r) + v_n\varepsilon_0(E_z^i + E_z^r)]_{x=vt} \\
&= \sqrt{\frac{\varepsilon_0}{\mu_0}}\left[\left(\cos\theta + \frac{v}{c_0}\right) + \left(\cos\tilde{\theta} - \frac{v}{c_0}\right)\frac{\sin\theta}{\sin\tilde{\theta}}\right] \\
&\qquad \Phi(y\sin\theta - (c_0 + v\cos\theta)t),
\end{aligned} \tag{5.53a}
$$

with (5.48b) being taken into account. The latter can also be written as

$$
J_{Sz} = \sqrt{\frac{\varepsilon_0}{\mu_0}}\left(1 - \frac{v^2}{c_0^2}\right)\frac{\sin(\theta + \tilde{\theta})}{\sin\tilde{\theta}}\Phi(y\sin\theta - (c_0 + v\cos\theta)t). \tag{5.53b}
$$

When $v = 0$ one gets $\tilde{\theta} = \theta$ and

$$
\begin{aligned}
J_{Sz} &= 2\sqrt{\frac{\varepsilon_0}{\mu_0}}\Phi(y\sin\theta - c_0 t) \\
&= 2H_y^i|_S,
\end{aligned} \tag{5.54}
$$

as expected while in the case of $\theta = 0$ (i.e., when the incident wave is normal to the plate), one has

$$
\begin{aligned}
J_{Sz} &= 2\sqrt{\frac{\varepsilon_0}{\mu_0}}\left(1 + \frac{v}{c_0}\right)\Phi(-(c_0 + v)t) \\
&= \left(1 + \frac{v}{c_0}\right)2H_y^i|_S.
\end{aligned} \tag{5.55}
$$

From this, one concludes that the contribution of the velocity to the surface current is of the order of $O(v/c_0)$.

PROBLEMS

1. **(a)** Transform the given incident wave (5.45a,b) as well as the *proposed* reflected wave (5.47a,b) into a reference system attached to the plate.

(b) Show that the above-mentioned transformed expressions give also waves which propagate with velocity equal to c_0.

(c) Solve the boundary-value problem connected with the reflection phenomenon in question in the reference system attached to the plate. Show that the directions of propagations of the incident and reflected waves are symmetrical positions with respect to the normal vector of the plate.

(d) Transform the expressions found in (c) into the original reference system *Oxyz* and compare the resulting expressions with (5.49a–c).

2. Assume that a perfectly conducting planar sheet is moving in *vacuum* with a constant velocity v in a direction which is normal to itself and is illuminated by a linearly polarized plane wave whose magnetic field is parallel to the sheet (see Fig. 5.3).

(a) Find the expression of the reflected field. What are the propagation direction and the frequency of the reflected wave?

(b) Compute the densities of the surface charge and surface current excited on the plate.

(c) Plot the curves which show the variations of the reflection angle and frequency of the reflected wave with the velocity for various values of the incidence angle (Consider both positive and negative values of the velocity).

(d) Plot the curves which show the variations of the charge and current densities with the velocity for various values of the incidence angle.

B. Discontinuities of the Field Excited by a Moving Combined Sheet [75]

As a second example, consider the so-called *combined sheet* that consists of a combination of a simple and double sheets. This kind of a configuration, which had been considered first in Panicali [17] and then in Van Bladel [18] and Idemen [16] for a sheet at rest, may occur when charges of opposite signs and unequal densities are accumulated

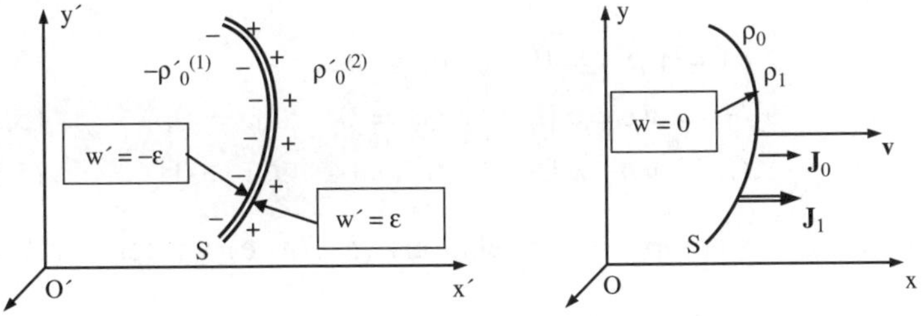

Figure 5.4. A moving combined sheet. **(a)** Seen from a co-moving reference system, **(b)** Seen from a rest system.

on different sides of S (see Fig. 5.4). If the densities of the charges located on the left and right sides of S are denoted in the co-moving reference system $O'x'y'z'$ by $[-\rho'^{(1)}_0]$ and $[\rho'^{(2)}_0]$, respectively, then the total charge density becomes

$$\begin{aligned}\rho' &= \lim_{\varepsilon\to 0}\{\rho'^{(2)}_0\delta(w'-\varepsilon) - \rho'^{(1)}_0\delta(w'+\varepsilon)\}\\ &= \lim_{\varepsilon\to 0}\{\rho'^{(1)}_0\delta(w'-\varepsilon) - \rho'^{(1)}_0\delta(w'+\varepsilon)\}\\ &\qquad + \lim_{\varepsilon\to 0}\{(\rho'^{(2)}_0 - \rho'^{(1)}_0)\delta(w'-\varepsilon)\}\\ &= -\lim_{\varepsilon\to 0}\{2\varepsilon\rho'^{(1)}_0\}\lim_{\varepsilon\to 0}\{[\delta(w'+\varepsilon)-\delta(w'-\varepsilon)]/(2\varepsilon)\\ &\qquad + (\rho'^{(2)}_0 - \rho'^{(1)}_0)\delta(w')\\ &= \rho'_1\delta'(w') + \rho'_0\delta(w'),\end{aligned} \tag{5.56a}$$

were we put

$$\rho'_0 = (\rho'^{(2)}_0 - \rho'^{(1)}_0), \qquad \rho'_1 = -\lim_{\varepsilon\to 0}\{2\varepsilon\rho'^{(1)}_0\}. \tag{5.56b}$$

From the latter, one concludes that the charge configuration in question consists of a combination of a simple sheet of density $\rho'_0 = (\rho'^{(2)}_0 - \rho'^{(1)}_0)$ with a double sheet of density $\rho'_1 = \{2\varepsilon\rho'^{(1)}_0\}$. In what follows we will assume that $[\rho'^{(1)}_0]$ and $[\rho'^{(2)}_0]$ are *independent* of t' (= the time measured in $O'x'y'z'$) and, hence, $\mathbf{J}'_0$ and $\mathbf{J}'_1$ are naught.

Now assume that the above-mentioned combined sheet moves in the direction parallel to the x axis with constant velocity v. Then the densities of charges and currents to be observed from the reference system $Oxyz$ are obtained through the Lorentz formulas (5.17e), namely,

$$\rho_0 = \gamma\rho'_0 \neq 0, \qquad \rho_1 = \gamma\rho'_1 \neq 0, \tag{5.57a}$$

$$J_{0x} = v\rho_0 \neq 0, \qquad J_{0y} = 0, \quad J_{0z} = 0, \tag{5.57b}$$

$$J_{1x} = v\rho_1 \neq 0, \qquad J_{1y} = 0, \quad J_{1z} = 0. \tag{5.57c}$$

In this case the compatibility relations (5.27a–e) written for $k = 2$ yield

$$\mathbf{n}\times\mathbf{H}_{1t} + v_n\varepsilon_0\mathbf{E}_{1t} = 0, \quad \mathbf{n}\times\mathbf{E}_{1t} - v_n\mu_0\mathbf{H}_{1t} = 0, \tag{5.57d}$$

$$E_{1n} = 0, \qquad H_{1n} = 0, \qquad J_{1n} = v_n\rho_1. \tag{5.57d'}$$

Here the sub-indices t and n stand for the tangential and normal components, respectively. By cross-multiplication of the first equation in (5.57d) by $\mathbf{n}$, one gets

$$v_n \varepsilon_0 \mathbf{n} \times \mathbf{E}_{1t} - \mathbf{H}_{1t} = 0,$$

which, together with second equation in (5.57d), yields

$$\mathbf{E}_{lt} = 0, \qquad \mathbf{H}_{1t} = 0 \tag{5.57e}$$

because one always has

$$\begin{vmatrix} v_n \varepsilon_0 & -1 \\ 1 & -v_n \mu_0 \end{vmatrix} = 1 - \left(\frac{v_n}{c_0}\right)^2 \neq 0.$$

From the remaining equations in (5.57d,d′) and (5.57e), one gets finally $\mathbf{E}_1 = \mathbf{H}_1 = 0$.

Similarly, by using (5.27a–e) written for $k \geq 2$, one can easily check that

$$\mathbf{E}_k = 0, \qquad \mathbf{H}_k = 0, \qquad k = 1, 2, \ldots. \tag{5.58}$$

By using (5.58) in the compatibility relations (5.27a–e) for $k = 1$, one writes also

$$\lambda \mathbf{n} \times \mathbf{H}_{0t} + \lambda v_n \varepsilon_0 \mathbf{E}_0 = \mathbf{J}_1, \qquad \lambda \mathbf{n} \times \mathbf{E}_{0t} - \lambda v_n \mu_0 \mathbf{H}_0 = 0, \tag{5.59a,b}$$

$$\lambda \varepsilon_0 \mathbf{E}_{0n} = \rho_1, \qquad H_{0n} = 0, \tag{5.59c,d}$$

and

$$\operatorname{div}\mathbf{J}_1 + \lambda J_{0n} + \frac{\partial}{\partial t}\rho_1 - \lambda v_n \rho_0 = 0. \tag{5.59e}$$

From (5.59a,b) one gets

$$E_{0n} = \frac{1}{v_n \varepsilon_0} J_{1n}/\lambda, \qquad \mathbf{E}_{0t} = -\frac{v_n \mu_0}{1 - (v_n/c_0)^2} \mathbf{J}_{1t}/\lambda \tag{5.60a,b}$$

$$H_{0n} = 0, \qquad \mathbf{H}_{0t} = -\frac{1}{1 - (v_n/c_0)^2} \mathbf{n} \times \mathbf{J}_1/\lambda. \tag{5.60c,d}$$

It is worthwhile to note that the compatibility relation (5.59e) is automatically satisfied by ρ_0, ρ_1, J_{0n} and $\mathbf{J}_1$ given by (5.57a–c) (see Problem 2 below).

Now consider the boundary relations (5.26a–d). They are as follows:

$$\lambda[[\mathbf{n} \times \mathbf{H}_t]] + \lambda v_n \varepsilon_0[[\mathbf{E}]]$$
$$= \operatorname{curl}\left\{\frac{1}{1-(v_n/c_0)^2}\mathbf{n} \times \mathbf{J}_1/\lambda\right\}$$
$$+ \frac{\partial}{\partial t}\left\{\frac{1}{v_n}J_{1n}/\lambda\mathbf{n} - \frac{v_n/c_0^2}{1-(v_n/c_0)^2}\mathbf{J}_{1t}/\lambda\right\} + \rho_0\mathbf{v} \tag{5.61a}$$

$$\lambda\varepsilon_0[[E_n]] = \rho_0 - \operatorname{div}\left\{\frac{1}{v_n}J_{1n}/\lambda\mathbf{n} - \frac{v_n/c_0^2}{1-(v_n/c_0)^2}\mathbf{J}_{1t}/\lambda\right\} \tag{5.61b}$$

$$\lambda[[\mathbf{n} \times \mathbf{E}_t]] - \lambda v_n \mu_0[[\mathbf{H}]]$$
$$= -\operatorname{curl}\left\{\frac{1}{v_n\varepsilon_0}J_{1n}/\lambda\mathbf{n} - \frac{v_n\mu_0}{1-(v_n/c_0)^2}\mathbf{J}_{1t}/\lambda\right\}$$
$$+ \frac{\partial}{\partial t}\left\{\frac{\mu_0}{1-(v_n/c_0)^2}\mathbf{n} \times \mathbf{J}_1/\lambda\right\} \tag{5.61c}$$

$$\lambda[[H_n]] = \operatorname{div}\left\{\frac{\mu_0}{1-(v_n/c_0)^2}\mathbf{n} \times \mathbf{J}_1/\lambda\right\} \tag{5.61d}$$

$$\lambda[[J_n]] - \lambda v_n[[\rho]] = -\frac{\partial}{\partial t}\rho_0. \tag{5.61e}$$

PROBLEMS

1. Solve the system (5.59a,b) and get (5.60a–d).
2. Show that the compatibility relation (5.59e) is automatically satisfied by ρ_0, ρ_1, J_{0n}, and $\mathbf{J}_1$ given by (5.57a–c)

5.3 DISCONTINUITIES ON A NONUNIFORMLY MOVING SHEET

Now consider the nonuniform motion of the boundary. In such a case we cannot use the formulas of the Special Theory of Relativity. However, the general formulas derived in Chapter 3 are still valid whenever the shape function $w(x, y, z, t)$ is known. In what follows we will consider some particular cases separately:

i. The case of a plane which moves in a direction normal to itself

ii. Plane interface between two simple media

iii. Vibrating nonperfectly conducting half-space

iv. Vibrating perfectly conducting half-space

v. Surface charges and currents on a plane

vi. Vibrating cone

5.3.1 Boundary Conditions on a Plane that Moves in a Direction Normal to Itself

Consider now the case when S consists of a plane and moves in a direction normal to itself with a velocity $\mathbf{v} = \mathbf{v}(t)$. Without loss of generality we can assume that $\mathbf{v}$ is parallel to the x axis. Then the function $w(x, y, x, t)$ mentioned in Section 3.2 and the resulting quantities λ, $\mathbf{n}$, and v_n become

$$w = x - \int^{t} v(\tau)\, d\tau, \qquad \mathrm{grad} w = \mathbf{n} = \mathbf{e}_x, \qquad \lambda = 1, \qquad v_n = v(t). \tag{5.62}$$

Thus (3.12a–f) and (3.13a–f) are reduced to

$$[[\mathbf{e}_x \times \mathbf{H}]] + v(t)[[\mathbf{D}]] = -\mathrm{curl}\mathbf{H}_0 + \frac{\partial}{\partial t}\mathbf{D}_0 + \mathbf{J}_0, \tag{5.63a}$$

$$[[\mathbf{e}_x \times \mathbf{E}]] - v(t)[[\mathbf{B}]] = -\mathrm{curl}\mathbf{E}_0 - \frac{\partial}{\partial t}\mathbf{B}_0, \tag{5.63b}$$

$$[[D_x]] = \rho_0 - \mathrm{div}\mathbf{D}_0, \tag{5.63c}$$

$$[[B_x]] = -\mathrm{div}\mathbf{B}_0, \tag{5.63d}$$

$$[[J_x]] - v(t)[[\rho]] = -\frac{\partial}{\partial t}\rho_0 - \mathrm{div}\mathbf{J}_0, \tag{5.63e}$$

$k \geq 1 \Rightarrow$

$$\mathrm{curl}\mathbf{H}_k + \mathbf{e}_x \times \mathbf{H}_{k-1} - \frac{\partial}{\partial t}\mathbf{D}_k + v(t)\mathbf{D}_{k-1} = \mathbf{J}_k, \tag{5.64a}$$

$$\mathrm{curl}\mathbf{E}_k + \mathbf{e}_x \times \mathbf{E}_{k-1} + \frac{\partial}{\partial t}\mathbf{B}_k - v(t)\mathbf{B}_{k-1} = 0, \tag{5.64b}$$

$$\mathrm{div}\mathbf{D}_k + D_{k-1,x} = \rho_k, \tag{5.64c}$$

$$\mathrm{div}\mathbf{B}_k + B_{k-1,x} = 0, \tag{5.64d}$$

$$\mathrm{div}\mathbf{J}_k + J_{k-1,x} + \frac{\partial}{\partial t}\rho_k - v(t)\rho_{k-1} = 0. \tag{5.64e}$$

Here the sub-indices (x) appearing in different terms signify the components parallel to the x axis. Equations (5.63a–e) are the most general form of the boundary conditions on the plane S while (5.64a–e) are the compatibility relations connected with the singular parts of the source and field functions. In what follows we will consider some more particular cases that will have important practical applications.

A. The Case of Simple Planar Interface Between Two Simple Media

In the case when S consists of a simple interface between two simple media, S has no physical existence, which yields for all $k = 0, 1, \ldots$.

$$\mathbf{P}_k = 0, \qquad \mathbf{M}_k = 0, \qquad \mathbf{D}_k = \varepsilon_0 \mathbf{E}_k, \qquad \mathbf{B}_k = \mu_0 \mathbf{H}_k. \tag{5.65a}$$

Owing to the possible difference between the conductances of the materials filling the regions taking place in different sides of S, one can observe an accumulation of charge, with density ρ_S, and a flow of surface current with density $\mathbf{J}_S$, on the interface. That means that all the charge and current components concentrated on S but ρ_0 and $\mathbf{J}_0$ are naught:

$$\rho_k = 0, \qquad \mathbf{J}_k = 0, \qquad k = 1, \ldots. \tag{5.65b}$$

Now assume that for $k \geq$ n+1 one has $\mathbf{E}_k = 0$ and $\mathbf{H}_k = 0$. Then from (5.64a–e) one writes first

$$\mathbf{e}_x \times \mathbf{H}_{n,t} + v(t)\varepsilon_0 \mathbf{E}_{n,t} = 0, \tag{5.66a}$$

$$\mathbf{e}_x \times \mathbf{E}_{n,t} - v(t)\mu_0 \mathbf{H}_{n,t} = 0, \tag{5.66b}$$

$$\varepsilon_0 \mathbf{E}_{n,x} = 0, \tag{5.66c}$$

$$\mu_0 \mathbf{H}_{n,x} = 0, \tag{5.66d}$$

$$J_{n,x} - v(t)\rho_n = 0, \tag{5.66e}$$

Here the sub-index t refers to the projection on the plane S. If we consider the cross-multiplication of (5.66b) with $\mathbf{e}_x$, then we see that

$$\mathbf{E}_{n,t} + v(t)\mu_0 \mathbf{e}_x \times \mathbf{H}_{n,t} = 0, \tag{5.67a}$$

which, together with (5.66a), yields

$$\mathbf{E}_{n,t} = 0, \qquad \mathbf{H}_{n,t} = 0 \tag{5.67b}$$

because, in accordance with the special theory of relativity, one always has

$$\begin{vmatrix} 1 & v(t)e_0 \\ v(t)\mu_0 & 1 \end{vmatrix} = 1 - \frac{v(t)^2}{c_0^2} > 0. \tag{5.68}$$

Repeating the above reasoning by replacing n by $(n-1)$, we convince ourselves that all the singular parts of the field components are naught; that is,

$$\mathbf{E}_k = 0, \qquad \mathbf{H}_k = 0, \qquad k = 0, 1, \ldots. \tag{5.69}$$

Thus, (5.63a–e) and (5.64e) are reduced to the following:

$$[[\mathbf{e}_x \times \mathbf{H}_t]] + v(t)[[\mathbf{D}]] = \mathbf{J}_0, \tag{5.70a}$$

$$[[\mathbf{e}_x \times \mathbf{E}_t]] - v(t)[[\mathbf{B}]] = 0, \tag{5.70b}$$

$$[[D_x]] = \rho_0, \tag{5.70c}$$

$$[[B_x]] = 0, \tag{5.70d}$$

$$[[J_x]] - v(t)[[\rho]] = -\frac{\partial}{\partial t}\rho_0 - \mathrm{div}\mathbf{J}_0 \tag{5.70e}$$

and

$$J_{x,0} = v(t)\rho_0. \tag{5.71}$$

Equations (5.70a–e) are the boundary conditions to be observed on a plane moving in a direction normal to itself.

B. A Vibrating Nonperfectly Conducting Half-Space

Let the region $x < 0$ be filled with a simple (*not perfectly conducting*) material while the region $x > 0$ consists of the vacuum. If at an instant, say at $t = 0$, the half-space begins to vibrate about its original position with a frequency equal to ω, which yields

$$w = x - a\sin(\omega t), \qquad v(t) = a\omega\cos(\omega t), \qquad t > 0. \tag{5.72}$$

In this case (5.63a–e) are replaced by

$$[[\mathbf{e}_x \times \mathbf{H}_t]] + a\omega\cos(\omega t)[[\mathbf{D}]] = \mathbf{J}_0, \tag{5.73a}$$

$$[[\mathbf{e}_x \times \mathbf{E}_t]] - a\omega\cos(\omega t)[[\mathbf{B}]] = 0, \tag{5.73b}$$

$$[[D_x]] = \rho_0, \tag{5.73c}$$

$$[[B_x]] = 0, \tag{5.73d}$$

$$[[J_x]] - a\omega\cos(\omega t)[[\rho]] = -\frac{\partial}{\partial t}\rho_0. \tag{5.73e}$$

C. A Vibrating Perfectly Conducting Half-Space

Assume now that the region $x < 0$ is filled with *perfectly conducting* material while the region $x > 0$ consists of the vacuum. Then (5.70a)–(5.71) are reduced to

$$\mathbf{e}_x \times \mathbf{H}_t^{(2)} + v(t)\varepsilon_0\mathbf{E}_t^{(2)} = \mathbf{J}_{0,t} \tag{5.74a}$$

$$\mathbf{e}_x \times \mathbf{E}_t^{(2)} - v(t)\mu_0\mathbf{H}_t^{(2)} = 0 \tag{5.74b}$$

$$\varepsilon_0 E_x^{(2)} = \rho_0 \tag{5.74c}$$

$$H_x^{(2)} = 0 \tag{5.74d}$$

$$J_x^{(2)} - v(t)\rho = -\frac{\partial}{\partial t}\rho_0 - \mathrm{div}\mathbf{J}_0, \tag{5.74e}$$

and

$$J_{0,x} = v(t)\rho_0. \tag{5.75}$$

From (5.74a) and (5.74b) we obtain

$$\mathbf{E}_t^{(2)} = -\frac{v(t)\mu_0}{1 - v(t)^2/c_0^2}\mathbf{J}_{0,t} \tag{5.76a}$$

and

$$\mathbf{H}_t^{(2)} = -\frac{1}{1 - v(t)^2/c_0^2}\mathbf{e}_x \times \mathbf{J}_{0,t}. \tag{5.76b}$$

The unexpected result stated by (5.76a) can be *interpreted as the existence of a magnetic current* concentrated on the surface S. Its density is proportional both to the velocity and the electric current density. The existence of the surface electric current causes also to the existence of the surface magnetic current.

D. Boundary Conditions for the Field Excited by Surface Charges and Currents Concentrated on a Moving Plane Located in a Simple Medium

Assume that certain surface charges and currents are concentrated on S while S is moving as explained in Section 5.3 above. In this case the formulas (5.63a)–(5.63e) are all also reduced to (5.70a)–(5.70e) with known $\rho_0, \mathbf{J}_0$, and $v(t)$ (one has also $\mathbf{D} = \varepsilon_0 \mathbf{E}$ and $\mathbf{B} = \mu_0 \mathbf{H}$). Thus one writes

$$[[\mathbf{e}_x \times \mathbf{H}]] + v(t)\varepsilon_0[[\mathbf{E}_t]] = \mathbf{J}_{0,t}, \tag{5.77a}$$

$$[[\mathbf{e}_x \times \mathbf{E}]] - v(t)\mu_0[[\mathbf{H}_t]] = 0, \tag{5.77b}$$

$$[[\mathrm{E}_x]] = \rho_0/\varepsilon_0, \tag{5.77c}$$

$$[[\mathrm{H}_x]] = 0, \tag{5.77d}$$

$$[[\mathrm{J}_x]] - v(t)[[\rho]] = -\frac{\partial}{\partial t}\rho_0 - \mathrm{div}\mathbf{J}_0, \tag{5.77e}$$

and

$$J_{0,x} = v(t)\rho_0. \tag{5.78}$$

From (5.77a,b) one gets

$$[[\mathbf{E}_t]] = -\frac{\mu_0 v(t)}{1 - v(t)^2/c_0^2}\mathbf{J}_{0,t} \tag{5.79a}$$

and

$$[[\mathbf{e}_x \times \mathbf{H}_t]] = \frac{1}{1 - v(t)^2/c_0^2}\mathbf{J}_{0,t}. \tag{5.79b}$$

5.3.2 Boundary Conditions on the Interface of Two Simple Media

Suppose that two simple media are in contact across a regular interface S that moves in accordance with (3.5). In this case one can observe only an accumulation of charge, say ρ_S, and a surface current, say $\mathbf{J}_S$, on the interface. That means that all the quantities concentrated on S but $\rho_0, \mathbf{J}_0, \mathbf{E}_0, \mathbf{H}_0, \mathbf{D}_0$ and $\mathbf{B}_0$ are naught, which reduces the boundary conditions (3.12a–f) and compatibility equations (3.25a–f) into the following:

$$\lambda[[\mathbf{n} \times \mathbf{H}]] + \lambda v_n[[\mathbf{D}]] = -\mathrm{curl}\mathbf{H}_0 + \frac{\partial}{\partial t}\mathbf{D}_0 + \mathbf{J}_0, \tag{5.80a}$$

$$\lambda[[\mathbf{n} \times \mathbf{E}]] - \lambda v_n[[\mathbf{B}]] = -\mathrm{curl}\mathbf{E}_0 - \frac{\partial}{\partial t}\mathbf{B}_0, \tag{5.80b}$$

$$\lambda[[\mathbf{n} \cdot \mathbf{D}]] = \rho_0 - \mathrm{div}\mathbf{D}_0, \tag{5.80c}$$

$$\lambda[[\mathbf{n}\cdot\mathbf{B}]] = -\mathrm{div}\mathbf{B}_0, \tag{5.80d}$$

$$\lambda[[\mathbf{n}\cdot\mathbf{J}]] - \lambda v_n[[\rho]] = -\frac{\partial}{\partial t}\rho_0 - \mathrm{div}\mathbf{J}_0, \tag{5.80e}$$

and

$$\lambda\mathbf{n}\times\mathbf{H}_0 + \lambda v_n\mathbf{D}_0 = 0, \tag{5.81a}$$

$$\lambda\mathbf{n}\times\mathbf{E}_0 - \lambda v_n\mathbf{B}_0 = 0, \tag{5.81b}$$

$$\lambda\mathbf{n}\cdot\mathbf{D}_0 = 0, \tag{5.81c}$$

$$\lambda\mathbf{n}\cdot\mathbf{B}_0 = 0, \tag{5.81d}$$

$$\lambda\mathbf{n}\cdot\mathbf{J}_0 - \lambda v_n\rho_0 = 0. \tag{5.81e}$$

By considering the identities $\mathbf{P}_0 \equiv 0$ and $\mathbf{M}_0 \equiv 0$, one writes

$$\mathbf{D}_0 = \varepsilon_0\mathbf{E}_0, \qquad \mathbf{B}_0 = \mu_0\mathbf{H}_0, \tag{5.82}$$

which reduces (5.81c,d) to

$$E_{0n} = 0, \qquad H_{0n} = 0. \tag{5.83}$$

Note that these latter are also compatible with (5.81a,b). If we use these results in (5.81a,b), we write for the tangential components

$$\mathbf{n}\times\mathbf{H}_{0t} + v_n\varepsilon_0\mathbf{E}_{0t} = 0 \tag{5.84a}$$

$$\mathbf{n}\times\mathbf{E}_{0t} - v_n\mu_0\mathbf{H}_{0t} = 0. \tag{5.84b}$$

By cross-multiplication of (5.84b) with $\mathbf{n}$, one obtains

$$\mathbf{E}_{0t} + v_n\mu_0\mathbf{n}\times\mathbf{H}_{0t} = 0. \tag{5.84c}$$

The compatibility condition of the homogeneous linear equations in (5.84a) and (5.84c) requires

$$\begin{vmatrix} 1 & v_n\varepsilon_0 \\ v_n\mu_0 & 1 \end{vmatrix} = 0$$

or

$$(v_n)^2 = (c_0)^2. \tag{5.85}$$

Since one always has $v_n < c_0$, (5.85) never met. Therefore from (5.84a) and (5.84c) one gets

$$\mathbf{E}_{0t} = 0, \qquad \mathbf{H}_{0t} = 0, \tag{5.86}$$

which, along with (5.83) yield

$$\mathbf{E}_0 = 0, \qquad \mathbf{H}_0 = 0. \tag{5.87}$$

Now let us insert (5.87) into the boundary conditions (5.80a–e) and use the definitions (see (3.8b) and (3.10d))

$$\rho_0\delta(w) = \rho_0\frac{1}{\lambda}\delta(S) \equiv \rho_S\delta(S), \tag{5.88a}$$

and

$$\mathbf{J}_0\delta(w) = \mathbf{J}_0\frac{1}{\lambda}\delta(S) \equiv \mathbf{J}_S\delta(S), \tag{5.88b}$$

which yield

$$\rho_S = \frac{1}{\lambda}\rho_0, \qquad \mathbf{J}_S = \frac{1}{\lambda}\mathbf{J}_0. \tag{5.89}$$

The result becomes

$$[[\mathbf{n}\times\mathbf{H}]] + v_n[[\mathbf{D}_t]] = \mathbf{J}_{st}, \tag{5.90a}$$

$$[[\mathbf{n}\times\mathbf{E}]] - v_n[[\mathbf{B}_t]] = 0, \tag{5.90b}$$

$$[[\mathbf{n}\cdot\mathbf{D}]] = \rho_S, \tag{5.90c}$$

$$[[\mathbf{n}\cdot\mathbf{B}]] = 0 \tag{5.90d}$$

and

$$\begin{aligned}[[\mathbf{n}\cdot\mathbf{J}]] - v_n[[\rho]] &= -\frac{1}{\lambda}\frac{\partial}{\partial t}\rho_0 - \frac{1}{\lambda}\mathrm{div}\mathbf{J}_0\\ &\equiv -\frac{1}{\lambda}\frac{\partial}{\partial t}(\lambda\rho_S) - \frac{1}{\lambda}\mathrm{div}(\lambda\mathbf{J}_{St}).\end{aligned} \tag{5.90e}$$

Note that the boundary relations (5.90a–d) are equivalent to the relations given in Namias [39] (see also references 40–44).

From (5.89) and (5.81e) one gets also

$$J_{Sn} \equiv \frac{1}{\lambda}\mathrm{J}_{0n} = \frac{1}{\lambda}v_n\rho_0 = \rho_S v_n. \tag{5.91}$$

This shows that the surface current consists of the motion of the surface charge. Note that when neither of the regions separated by S is *perfectly conducting*, one has $\mathbf{J}_S = 0$, which yields also $\rho_S = 0$ (cf. (5.91)).

A. Vibrating Cone

Let (r, θ, ϕ) stand for the usual spherical coordinates through which the equation of S is written as

$$\theta = \theta_0 + a\sin\omega_1 t, \tag{5.92a}$$

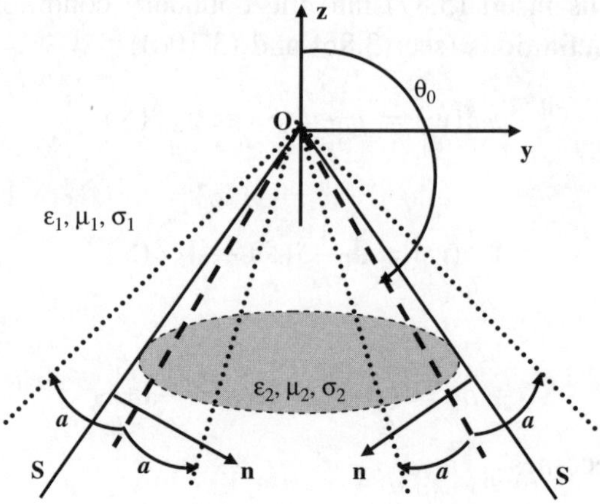

Figure 5.5. A sinusoidally vibrating cone.

where θ_0, a, and ω_1 show certain positive quantities (see Fig. 5.5) That means that the function $w(x, y, z, t)$ mentioned in (3.5) and the unit normal vector mentioned in (3.8) are now

$$w = \theta - \theta_0 - a \sin \omega_1 t, \qquad \mathbf{n} = \mathbf{e}_\theta \tag{5.92b}$$

Therefore (3.10c) gives

$$\frac{\partial}{\partial t} w = -(\mathbf{v} \cdot \mathbf{n})|\text{grad}\theta| = -v_n||\frac{\mathbf{e}_\theta}{r}| = -\frac{v_n}{r},$$

which yields

$$(\mathbf{v} \cdot \mathbf{n}) = -\mathrm{r}\frac{\partial w}{\partial t} = ar\omega_1 \cos \omega_1 t. \tag{5.92c}$$

Thus the apparent boundary conditions (5.90a–d) and (5.91) become

$$\mathbf{e}_\theta \times (\mathbf{H}_t^{(2)} - \mathbf{H}_1^{(1)}) + ar\omega_1 \cos \omega_1 t(\mathbf{D}_t^{(2)} - \mathbf{D}_1^{(1)}) = \mathbf{J}_{st}, \tag{5.93a}$$

$$\mathbf{e}_\theta \times (\mathbf{E}_t^{(2)} - \mathbf{E}_1^{(1)}) - ar\omega_1 \cos \omega_1 t(\mathbf{B}_t^{(2)} - \mathbf{B}_1^{(1)}) = 0, \tag{5.93b}$$

$$D_\theta^{(2)} - D_\theta^{(1)} = \rho_S, \tag{5.93c}$$

$$\mathrm{B}_\theta^{(2)} - B_\theta^{(1)} = 0, \tag{5.93d}$$

$$J_{s\theta} = ar\omega_1 \cos(\omega_1 t)\rho_S. \tag{5.93e}$$

Now let us discuss two important particular cases separately.

a. The Case When Both Regions Separated by S Are Not Perfectly Conducting. In such a case, $\mathbf{J}_S$ is naught on the interface,* which reduces (5.93a–d) into

$$\mathbf{e}_\theta \times (\mathbf{H}_t^{(2)} - \mathbf{H}_1^{(1)}) + ar\omega_1 \cos\omega_1 t \cdot (\varepsilon_2 E_t^{(2)} - \varepsilon_1 E_1^{(1)}) = 0, \tag{5.94a}$$

$$\mathbf{e}_\theta \times (\mathbf{E}_t^{(2)} - \mathbf{E}_1^{(1)}) - ar\omega_1 \cos\omega_1 t \cdot (\mu_2 \mathbf{H}_t^{(2)} - \mu_1 \mathbf{H}_1^{(1)}) = 0, \tag{5.94b}$$

$$\varepsilon_2 E_\theta^{(2)} - \varepsilon_1 E_\theta^{(1)} = \rho_S, \tag{5.94c}$$

$$\mu_2 H_\theta^{(2)} - \mu_1 H_\theta^{(1)} = 0, \tag{5.94d}$$

$$J_{s\theta} = ar\omega_1 \cos(\omega_1 t)\rho_S. \tag{5.94e}$$

These are the boundary conditions on the vibrating cone.

b. The Case When One of the Regions Separated by S is Perfectly Conducting. Assume without loss of generality that the region (2) is electrically perfectly conducting. In this case, ρ_S and $\mathbf{J}_S$ may be different from zero while one has $\mathbf{E}_t^{(2)} \equiv 0$ and $\mathbf{H}_t^{(2)} \equiv 0$. Thus (5.93a–e) are replaced by

$$\mathbf{e}_\theta \times \mathbf{H}_1^{(1)} + ar\omega_1 \cos\omega_1 t \cdot \varepsilon_1 E_1^{(1)} = -\mathbf{J}_{st}, \tag{5.95a}$$

$$\mathbf{e}_\theta \times \mathbf{E}_1^{(1)} - ar\omega_1 \cos\omega_1 t \cdot \mu_1 \mathbf{H}_1^{(1)} = 0, \tag{5.95b}$$

$$\varepsilon_1 \mathrm{E}_\theta^{(1)} = -\rho_S, \tag{5.95c}$$

$$H_\theta^{(1)} = 0, \tag{5.95d}$$

$$J_{s\theta} = ar\omega_1 \cos(\omega_1 t)\rho_S. \tag{5.95e}$$

Note that in this case the field in region 1 is determined by using only the Maxwell equations and the boundary conditions (5.95b,d). Then the densities of the accumulated surface charge ρ_S as well as the surface current $\mathbf{J}_{st}$ are found from (5.95a,c,e).

PROBLEMS

1. Let (r, ϕ, z) denote the usual cylindrical coordinates and let S be the half-plane oscillating *in the vacuum* about $\phi = \phi_0$ according to the law $\phi = \phi_0 + a\sin(\omega_1 t)$ (see the accompanying figure). Then show that
   (a) $w = \phi - \phi_0 - a\sin(\omega_1 t), \mathbf{n} = \mathbf{e}_\phi, \partial w/\partial t = -v_\phi/r = -a\omega_1 \cos(\omega_1 t)$.

*This claim is valid in the reference system moving with S. If the motion were uniform, then by the Lorentz transformation they would also be correct in the reference system $Oxyz$.

(b) The apparent boundary conditions on S are

$$\mathbf{e}_\phi \times (\mathbf{H}_t^{(2)} - \mathbf{H}_1^{(1)}) + ar\omega_1 \cos\omega_1 t \cdot (\mathbf{D}_t^{(2)} - \mathbf{D}_1^{(1)}) = \mathbf{J}_{st}$$
$$\mathbf{e}_\phi \times (\mathbf{E}_t^{(2)} - \mathbf{E}_1^{(1)}) - ar\omega_1 \cos\omega_1 t \cdot (\mathbf{B}_t^{(2)} - \mathbf{B}_1^{(1)}) = 0$$
$$D_\phi^{(2)} - D_\phi^{(1)} = \rho_S, \qquad B_\phi^{(2)} - B_\phi^{(1)} = 0$$
$$J_{s\phi} = ar\omega_1 \cos(\omega_1 t)\rho_S .$$

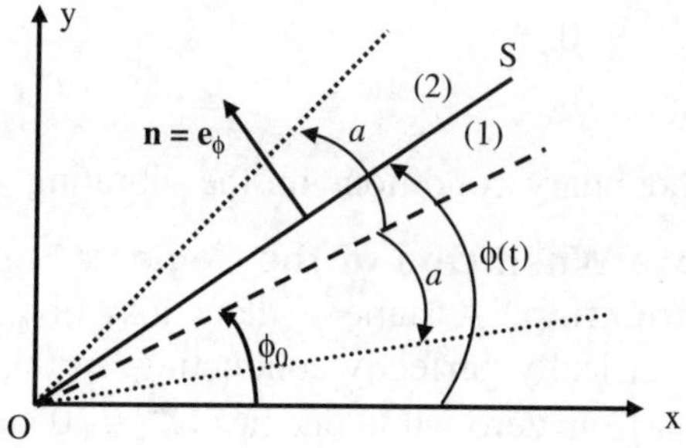

2. Reconsider the problem 1 in the case when S rotates about the z axis with a constant angular velocity ω_1, which yields $\phi = \omega_1 t$.
3. Let S be an oscillating plane that separates two simple media.

   (a) Assuming the equation of S is to be given by $z = a\sin(\omega_1 t)$, find the apparent boundary conditions on S.

   (b) Discuss the cases of $\sigma_1 < \infty$ and $\sigma_1 = \infty$.

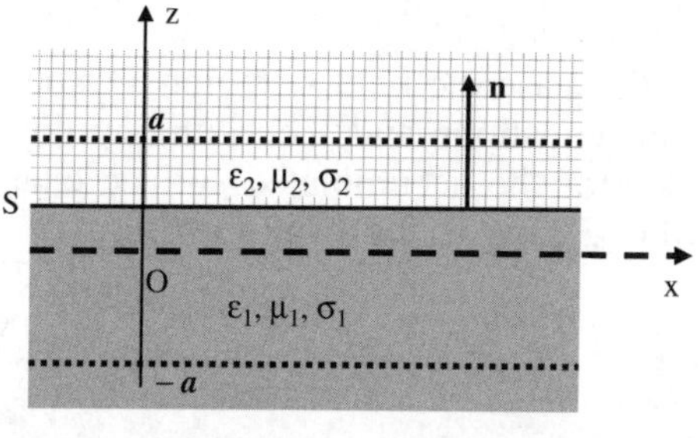

4. Reconsider problem 3 in the case when S makes a uniform motion with velocity v such that on S one has $z = vt$.

CHAPTER 6

Edge Singularities on Material Wedges Bounded by Plane Boundaries

6.1 INTRODUCTION

From both experimental and theoretical works it is known that some components of the electromagnetic field exhibit singularities of certain types near sharp edges (see Fig. 6.1). These singularities affect the performance of electric devices and, hence, require serious precautions at both the design and production stages. Therefore, one has to know beforehand the types as well as the orders of possible singularities. Furthermore, in theoretical investigations concerning structures involving edges, the singularities cause some hard mathematical difficulties. Their knowledge in analytical investigations (based, for example, on the Wiener–Hopf or Riemann–Hilbert techniques) permits one to determine some auxiliary functions while in numerical works it enables one to improve the convergence of some procedures if the singularities are incorporated into the field expressions explicitly. Hence, one has to try to establish a clear theory that is based on a sound basis and is as general as possible. The aim of the present chapter is to investigate these singularities for a wedge formed by *material* planes in a rather general form by simple but rigorous approach.

The history of the works on edge-type singularities for the electromagnetic field goes back to a work by Bouwkamp in 1946 [1].

Discontinuties in the Electromagnetic Field, First Edition. By M. Mithat Idemen.

Published 2011 by John Wiley & Sons, Inc.

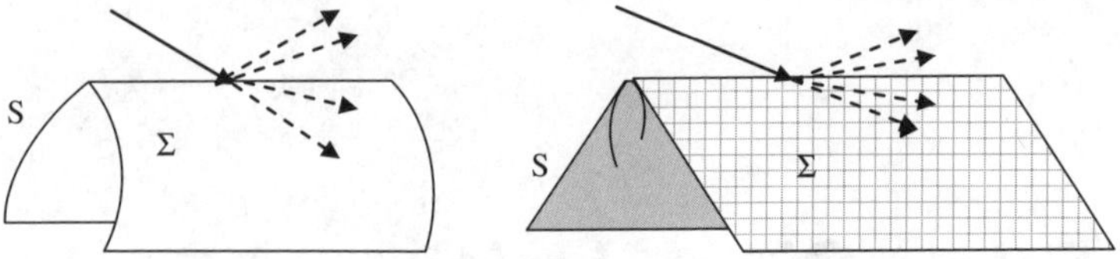

Figure 6.1. Edge-diffracted waves on a material sheet.

In this work connected with a perfectly conducting half-plane located in an infinite homogeneous space, he showed that it is possible to find many solutions satisfying both the Maxwell equations and the boundary conditions. They differ from each other by their singularities near the edge, which are, in general, not physically acceptable. If one knows the physically acceptable singularity, then one can find the unique, physically meaningful solution. The type and order of the edge singularity, which is admissible by physics laws, is referred to as the *edge conditions*.

The same problem was reconsidered three years later by Meixner [45], who postulated infinite series expansions of the following form for the Debye potentials:

$$\Pi_{\text{i}}(r,\phi) = A_i(\phi) + B_i(\phi)r + C_i((\phi)r^{3/2} + \cdots, \qquad i = 1,2. \quad (6.1)$$

Here (r, ϕ, z) stand for the circular cylindrical coordinates such that $r = 0$ corresponds to the edge. This *Ansatz* was later slightly modified by many investigators including Meixner himself so as to apply to various geometries (such as wedges) and boundary conditions [46–55]. A modification proposed by Meixner himself consisted of writing Frobenius-type infinite series, which were very successful in the theory of ordinary linear differential equations, for each component of the field, namely [52],

$$E_r(r,\phi) = a_0(\phi)r^{t-1} + a_1(\phi)r^t + a_2(\phi)r^{t+1} + \cdots. \quad (6.2)$$

(similar expressions for the other components). Here t stands for a number that will be determined through both the Maxwell equations and the boundary conditions. It is not the full expansions like (6.2) but the *minimum* value of all possible exponents t which is important for the present purpose.

The basic Ansatzes of Meixner, namely (6.1) and (6.2), were criticized by several investigators because

i. They had no sound basis and, consequently, although the first terms they yield are correct, their higher-order terms are invalid (nonexisting) [56, 57].

ii. They do not involve logarithmic singularities and, consequently, their results are sometimes wrong [58].

In order to establish a sound theory, some investigators proposed to use the results to be derived from the solution of the static field problems (electric and magnetic) connected with the same geometrical and physical conditions [56, 58]. One of the drawbacks of the *static field approach* is that they start from the Laplace equation and consider merely an electric (or a magnetic) field not coupled to a magnetic (or an electric) field whereas many boundary conditions that are actually in extensive use (for example, impedance and resistive-type conditions) involve electric and magnetic fields simultaneously [56].

To modify Meixner's Ansatz (6.2) so as to involve also the logarithmic terms, Braver et al. [57] proposed to write

$$\Psi(r,\phi) = r^{\tau} \sum_{m=0}^{\infty} \sum_{n=0}^{m} C_{mn}(\phi) r^{m} \ln^{n} r, \qquad \tau \geq 0. \tag{6.3}$$

Thus the terms such as $\ln r, (\ln r)^2$, and so on, which cannot be explored through Meixner's Ansatz, can now be discovered. It is obvious that the above-mentioned critiques directed to (6.2) can also be repeated for (6.3). For example, according to the well-known theory of the ordinary linear differential equations, near a regular-singular point the solution of an equation of the *second order* may involve only $\ln r$ but not $(\ln r)^n$ with $n \geq 2$. Secondly, from the recurrence relations to be satisfied by C_{mn} it is not easy to find the maximum power of the logarithms. Furthermore, it is rather hard to elaborate the expressions like (6.3) to derive results concerning rather complicated boundary conditions and geometrical configurations.

A different and seemingly more effective and sound approach that depends on the concept of *separable solutions* was used by Jones [3, p. 567]. In order to discover the singularities, he postulates an expansion of the form

$$\Pi = \sum \sum a_{m\nu} Y_{\nu}\{(k^2 - m^2)^{1/2} r\} e^{imz} \frac{\sin \nu\phi}{\cos \nu\phi}, \tag{6.4}$$

where Y_ν stands for the usual Neumann function which is singular when $r \to 0$. Thus he derives the same results that can be obtained by the methods cited above. In Idemen [59] the concept of separable solutions was combined with the concept of *confluence* so as to establish a sound and general method that permits one to reveal also logarithmic singularities, if any.

Finally, to conclude this short historical note on the edge-singularity, we want also to draw the attention to some methods depending on integral equations (see Maue [60] and Jones [61]). They are much more general as compared to the above-mentioned ones and allow directly for curved edges as well as those that change direction discontinuously. However, in a situation that concerns rather general boundary conditions, they seem to result in too complicated equations.

In the present chapter one considers the edges on a wedge with plane boundaries. By considering the diversity of the boundary conditions that frequently appear in the open literature as well as those that seem to have important practical applications in near future, we will try to formulate the problem in its most general form. To this end, we will assume that the faces of the wedge are formed by arbitrary material sheets and apply the above-mentioned method of *separable solutions* combined with the concept of *confluence*, which will both produce the logarithmic singularity, if any, and clarify the reason behind this type of singularity.

In what follows, one distinguishes two main groups according to whether the boundary is penetrable or not. The case where the boundary of the wedge is penetrable is treated in Section 6.3, while the case of impenetrable wedge is considered in Section 6.4. In Section 6.5 we apply the results to a particular case where the wedge consists of a half-plane with different face properties. In Section 6.6 we give a short discussion of the edge conditions for the induced surface currents, which play important roles in some theoretical investigations.

In this section we will consider only monochromatic waves with time dependence $\exp(-i\omega t)$ and omit both the symbol $\Re$ (real part) and the factor $\exp(-i\omega t)$.

6.2 SINGULARITIES AT THE EDGES OF MATERIAL WEDGES

Since the behavior we want to derive concerns only an infinitely small neighborhood of a fixed point on the edge, without loss of generality we can assume that the wedge is bounded by two half-planes (see Fig. 6.2). If the faces of the wedge are not exactly plane, then an equivalent wedge with plane boundaries can be formed by considering the tangent planes at the given point. For the sake of mathematical simplicity, we will define a coordinate system $Oxyz$ such that the wedge is symmetrical with respect to the plane $y = 0$ and the edge is coincident with the z axis. We will also assume that the given point is the origin and the field does not depend on the z coordinate.

The faces S and Σ of the wedge are supposed to be *material* half-planes that may, in general, be *penetrable*. Hence, one has to consider both regions B and B^* separated by the wedge. Then the limiting field values on the boundary S, observed from sides 1 and 2, are interrelated by two boundary conditions that can be written in the following forms (see (4.26a,b)):

$$a_1\mathbf{n}\times\mathbf{H}_{\mathrm{t}} + a_2\mathbf{n}\times\mathbf{H}_{\mathrm{t}}^* + a_3\mathbf{E}_t + a_4\mathbf{E}_t^* + a_5\mathbf{n}\times\mathrm{grad}H_n + a_6\mathbf{n}\times\mathrm{grad}\mathbf{H}_n^* + a_7\mathrm{grad}E_n + a_8\mathrm{grad}E_n^* = 0 \tag{6.5a}$$

$$b_1\mathbf{n}\times\mathbf{H}_t + b_2\mathbf{n}\times\mathbf{H}_{\mathrm{t}}^* + \cdots\cdots \quad \cdots\cdots + b_8\mathrm{grad}E_n^* = 0. \tag{6.5b}$$

In these equations, $\mathbf{n}$ stands for the unit normal on S directed from side 1 to side 2 while the sub-indices n and t refer to the normal and tangential components, respectively. As to the super-index (*), it refers to the values observed inside region B^*. Note that the values observed inside the region B have no special indications. This

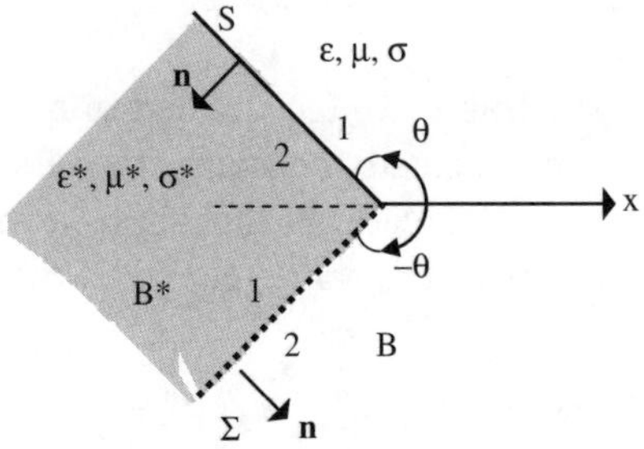

Figure 6.2. A wedge bounded by material sheets.

notation will largely be used in what follows to indicate the quantities related to or defined in the regions B and B^*. The coefficients a_i and $b_i (i = 1, 2, \ldots, 8)$ appearing in (6.5a) and (6.5b) are known complex constants that can be determined through the constitutive parameters of the material of the sheet modeled by S (see (4.25a–d)). If the material of the sheet is isotropic in tangential directions (it may be anisotropic when both the tangential and the normal directions are considered together), then (6.5a) and (6.5b) can model the sheet S rather accurately (see (4.11a–4.15e)). Similar relations are also written for the face Σ. In what follows, the coefficients related to Σ will be denoted by α_i and $\beta_i (i = 1, 2, \ldots, 8)$, namely,

$$\alpha_1 \mathbf{n} \times \mathbf{H}_t^* + \alpha_2 \mathbf{n} \times \mathbf{H}_t + \alpha_3 \mathbf{E}_t^* + \alpha_4 \mathbf{E_t} + \alpha_5 \mathbf{n} \times \mathrm{grad} H_n^* + \alpha_6 \mathbf{n} \times \ \mathrm{grad} H_n + \alpha_7 \ \mathrm{grad} E_n^* + \alpha_8 \ \mathrm{grad} E_n = 0 \quad (6.6a)$$

$$\beta_1 \mathbf{n} \times \mathbf{H}_t^* + \beta_2 \mathbf{n} \times \mathbf{H}_t + \cdots\cdots\cdots \ \cdots\cdots + \beta_8 \ \mathrm{grad} E_n = 0. \quad (6.6b)$$

If all the coefficients with even indices in (6.5a) and odd indices in (6.5b) are zero (or (6.5a) and (6.5b) can be reduced to such a system), then S constitutes an impenetrable boundary. The same is true for (6.6a), (6.6b), and Σ also. When both S and Σ are impenetrable, the boundary $S + \Sigma$ becomes a shield and separates the regions B and B^* completely from each other. In such a case the regions B and B^* have to be considered separately. In what follows, we will first consider the case where the boundary $S + \Sigma$ is penetrable. Note that in this case S or Σ alone or both of them can be penetrable. The results concerning the impenetrable wedge will then be derived from those concerning the penetrable one by straightforward simplifications.

6.3 THE WEDGE WITH PENETRABLE BOUNDARIES

Consider now the penetrable wedge defined above and suppose that the exciting sources are located far away from the edge, which is a plausible assumption. In this case all the Cartesian components of the electromagnetic field satisfy the same reduced wave equation near the edge, namely,

$$\Delta u(r, \phi) + k^2 u(r, \phi) = 0, \qquad r \in [0, r_0), (r, \phi) \in B \qquad (6.7a)$$

$$\Delta u(r, \phi) + k^{*2} u(r, \phi) = 0, \qquad r \in [0, r_0), (r, \phi) \in B^*, \qquad (6.7b)$$

where r and ϕ are the usual cylindrical polar coordinates while k and k^* stand for the wave numbers of the regions B and B^*, respectively. As to r_0, it is a certain positive number determined by the locations of the sources.

As is well known, every particular solution to (6.7a), which satisfies certain boundary conditions, consists, in general, of discrete or continuous combinations of its *separable solutions* [62]. These latter are of the form $u = R_\nu(r)\Phi_\nu(\phi)$ with [63]

$$R_\nu(r) = \begin{cases} J_\nu(kr) & \text{when } \nu \notin \{0, \pm 1, \pm 2, \ldots.\}, \\ J_n(kr) + A_n Y_n(kr) & \text{when } \nu = n \in \{0, 1, 2, \ldots\} \end{cases} \tag{6.8a}$$

and

$$\Phi_\nu(\phi) = \begin{cases} B_\nu \cos \nu\phi + C_\nu \sin \nu\phi & \text{when } \nu \neq 0, \\ B_0 + C_0\phi & \text{when } \nu = 0. \end{cases} \tag{6.8b}$$

Here ν stands for the separation parameter while J_ν and Y_n denote the usual Bessel and Neumann functions. The set of ν that has to be taken into account to construct a particular solution

$$u(r, \phi) = \sum_\nu R_\nu(r)\Phi_\nu(\phi) \tag{6.9a}$$

or

$$u(r, \phi) = \int_\Sigma R_\nu(r)\Phi_\nu(\phi) d\nu \tag{6.9b}$$

is the *spectrum* of u. The spectrum as well as the constants A_n, B_ν, and C_ν appearing in (6.8a,b) can be determined by considering also the sources as well as the boundary and radiation conditions, which is beyond our present aim. Our aim consists of revealing the asymptotic behaviors of the field components when $r \to 0$. To this end, we have to consider the following expressions [12] of $J_\nu(\eta)$ and $Y_n(\eta)$, which are valid near $\eta = 0$:

$$J_\nu(\eta) = c_{\nu 1}\eta^\nu + c_{\nu 2}\eta^{\nu+2} + \cdots \qquad (\nu \neq -1, -2, \ldots), \tag{6.10a}$$

$$Y_n(\eta) = d_{n1}\eta^{-n} + d_{n2}\eta^{-n+2} + \cdots + [\gamma_{n1}\eta^n + \gamma_{n2}\eta^{n+2} + \cdots] \log \eta. \tag{6.10b}$$

Here $c_{\nu j}, d_{nj}$, and γ_{nj} stand for certain known numbers. From (6.10a,b) one concludes at first glance that when $r \to 0$ the field can become singular having infinitely large values as a negative power of r or logarithm of r.

To reveal the asymptotic behaviors (and admissible singularities) of the field components for $r \to 0$, we will (and must!) take into account the fact that:

i. The field components satisfy the Maxwell equations.

ii. The energy stored near the edge is finite.

iii. The boundary conditions on $S \cup \Sigma$ are satisfied.

In what follows we will study the contributions of these restrictions separately. In order to reduce useless mathematical complications, we will consider the E and H cases separately. In the E case the electric field is parallel to the edge while in the H case the magnetic field is parallel to it. Obviously this is not a severe restriction because by adding the results of these particular cases one gets the expressions concerning the general case.

6.3.1 The H Case

When the magnetic field is parallel to the edge, all the nonzero components of the electromagnetic field near the edge in B can be written in terms of a scalar potential u satisfying (6.7a), namely [63],

$$H_z = k^2 u, \qquad E_r = \frac{i\omega\mu}{r}\frac{\partial u}{\partial \phi}, \qquad E_\phi = -i\omega\mu\frac{\partial u}{\partial r}. \tag{6.11}$$

By replacing k and μ in (6.11) by k^* and μ^*, respectively, one gets the expressions valid in B^*. By considering these expressions with the restrictions i–iii mentioned above, we get the following results.

A. Contribution of the Energy Restriction

As is well known, the densities of the stored electric and magnetic energies are proportional to $|\mathbf{E}|^2$ and $|\mathbf{H}|^2$, respectively. In order for the total electromagnetic energy stored inside a finite cylinder *of unit height* including the edge to be finite, one has to have

$$\iint |\mathbf{E}|^2\, dv = \iint |\mathbf{E}|^2 r\, dr d\phi < \infty, \tag{6.12a}$$

$$\iint |\mathbf{H}|^2\, dv = \iint |\mathbf{E}|^2 r\, dr d\phi < \infty, \tag{6.12a'}$$

which yield

$$(r)^{1/2}H_z(r,.),\quad (r)^{1/2}E_r(r,.),\quad (r)^{1/2}E_\phi(r,.) \in L^2[0,r_0] \tag{6.12b}$$

or, more precisely,

$$(r)^{1/2}R_\nu,\quad (r)^{-1/2}R_\nu d\Phi_\nu/d\phi,\quad (r)^{1/2}dR_\nu/dr \in L^2[0,r_0], \tag{6.12c}$$

with (6.11) being taken into account. On the other hand, (6.8a,b) and (6.10a,b) yield

$$R_\nu(r) \sim \begin{cases} \text{constant} \cdot r^\nu & \text{when } \nu \neq n, \\ \text{constant} \cdot r^n + \text{constant} \cdot A_n r^{-n} & \text{when } \nu = \text{n} \geq 1, \\ 1 + \text{constant} \cdot A_0 \log r & \text{when } \nu = n = 0. \end{cases} \tag{6.13}$$

By using (6.13) and (6.8b) in (6.12c), one gets easily

$$\Re\nu > 0 \qquad \text{when } \nu \neq n, \tag{6.14a}$$

$$A_n = 0 \qquad \text{when } n = 0, 1, \ldots, \tag{6.14b}$$

and

$$C_0 = 0. \tag{6.14c}$$

Here $\Re\nu$ stands, as usual, for the real part of ν. From (6.14b) we conclude that the separable solutions cannot individually involve logarithmic singularities. Such a singularity in the field may come from a certain kind of *confluence* that generates *nonseparable* solutions (see Section 6.3.1.G below).

B. Contribution of the Maxwell Equations

Because of (6.14a–c), the above-mentioned potential function u consists of a combination of certain functions u_ν whose *dominant* asymptotic expressions are as follows:

$$u_\nu \sim \begin{cases} B_0 & \text{when } \nu = 0 \\ [B_\nu \cos \nu\phi + C_\nu \sin \nu\phi]\, r^\nu & \text{when } \Re\nu > 0. \end{cases} \tag{6.15a}$$

The parameter ν appearing above is the number which both belongs to the spectrum and has the *smallest positive* real part. All the numbers ν that will appear below will designate this minimal ν which is referred to as the *minimal exponent*. We suppose first that

$$\nu \neq 1, 2, \ldots. \tag{6.15b}$$

The case where $v = n \in \{1, 2, \ldots\}$ will then be treated as being a limiting case $v \to n$ as certain parameters approach some critical values (see the remark in Section 6.3.1.E below).

By using (6.15a,b) in (6.11) we get

$$E_r \sim r^{v-1}[P \cos v\phi + Q \sin v\phi], \tag{6.16a}$$

$$E_\phi \sim r^{v-1}[Q \cos v\phi - P \sin v\phi], \tag{6.16b}$$

$$H_z \sim k^2 B_0 + ik^2/(v\omega\mu) r^v [Q \cos v\phi - P \sin v\phi], \tag{6.16c}$$

where P, Q, and B_0 are certain constants to be determined through the boundary conditions.

Note that (6.16a–c) are valid inside B for which $\phi \in (-\theta, \theta)$ (see Fig. 6.2). The expressions related to the region B^*, for which $\phi \in (\theta, 2\pi - \theta)$, are quite similar, namely,

$$E_r \sim r^{v-1}[P^* \cos v\phi + Q^* \sin v\phi], \tag{6.17a}$$

$$E_\phi \sim r^{v-1}[Q^* \cos v\phi - P^* \sin v\phi], \tag{6.17b}$$

$$H_z \sim k^{*2} B_0^* + ik^{*2}/(v\omega\mu^*) r^v [Q^* \cos v\phi - P^* \sin v\phi]. \tag{6.17c}$$

From (6.16a)–(6.17c) it is obvious that when $\Re v > 1$, there is no singularity (of the *algebraic* type) in the field in question. In other words, from the algebraic singularity point of view it is the case of $\Re v < 1$ which is important.

C. Contribution of the Boundary Conditions

In order to reveal the relations satisfied by the minimal exponent v as well as the constants B_0 and B_0^*, we have to consider all (*possible*) terms in (6.9a) which, after being inserted into (6.5a)–(6.6b), *produce* powers of orders r^v, r^{v-1}, r^{v-2} and r^0. They are obviously the terms involving $J_0(kr), J_1(kr), J_2(kr), J_v(kr), J_{v+1}(kr)$, and $J_{v+2}(kr)$. Hence, without loss generality, we can assume that

$$\begin{aligned} u(r, \phi) = {} & B_0 J_0(kr) + J_1(kr)[B_1 \cos\phi + C_1 \sin\phi] \\ & + J_2(kr)[B_2 \cos 2\phi + C_2 \sin 2\phi] \\ & + J_v(kr)[B_v \cos v\phi + C_v \sin v\phi] \end{aligned}$$

$$+ J_{v+1}(kr)[B_{v+1}\cos(v+1)\phi + C_{v+1}\sin(v+1)\phi]$$
$$+ J_{v+2}(kr)[B_{v+2}\cos(v+2)\phi + C_{v+2}\sin(v+2)\phi] + \cdots. \tag{6.18a}$$

By considering (6.10a), we can first write $u(r,\phi)$ as

$$\begin{aligned}u(r,\phi) = {} & [c_{01} + c_{02}(kr)^2 + \cdots]B_0 \\ & + [c_{11}(kr) + c_{12}(kr)^3 + \cdots][B_1\cos\phi + C_1\sin\phi] \\ & + [c_{21}(kr)^2 + c_{22}(kr)^4 + \cdots][B_2\cos 2\phi + C_2\sin 2\phi] \\ & + [c_{v1}(kr)^v + c_{v2}(kr)^{v+2} + \cdots][B_v\cos v\phi + C_v\cos v\phi] \\ & + [c_{v+1,1}(kr)^{v+1} + c_{v+1,2}(kr)^{v+3} + \cdots][B_{v+1}\cos(v+1)\phi \\ & + C_{v+1}\cos(v+1)\phi] \\ & + [c_{v+2,1}(kr)^{v+2} + c_{v+2,2}(kr)^{v+4} + \cdots][B_{v+2}\cos(v+2)\phi \\ & + C_{v+2}\cos(v+2)\phi] + \cdots\end{aligned} \tag{6.18b}$$

and then rearrange it as an ascending power series of r, namely,

$$\begin{aligned}u(r,\phi) = {} & c_{01}B_0 + c_{11}[B_1\cos\phi + C_1\sin\phi](kr) \\ & + \{c_{02}B_0 + c_{21}[B_2\cos 2\phi + C_2\sin 2\phi]\}(kr)^2 \\ & + c_{v1}[B_v\cos v\phi + C_v\sin v\phi](kr)^v + c_{v+1,1}[B_{v+1}\cos(v+1)\phi \\ & + C_{v+1}\sin(v+1)\phi](kr)^{v+1} \\ & + \{c_{v2}[B_v\cos v\phi + C_v\sin v\phi] \\ & + c_{v+2,1}[B_{v+2}\cos(v+2)\phi + C_{v+2}\sin(v+2)\phi]\}(kr)^{v+2} + \cdots.\end{aligned} \tag{6.18c}$$

If these expressions of $u(r,\phi)$ is used in (6.11), then one gets

$$\begin{aligned}H_z = {} & k^2c_{01}B_0 + k^2c_{11}[B_1\cos\phi + C_1\sin\phi](\mathrm{kr}) \\ & + k^2c_{v1}[B_v\cos v\phi + C_v\sin v\phi](kr)^v + \cdots,\end{aligned} \tag{6.19a}$$

$$\begin{aligned}E_r = {} & i\omega\mu c_{11}k[-B_1\sin\phi + C_1\cos\phi] + 2i\omega\mu c_{21}[-B_2\sin 2\phi \\ & + C_2\cos 2\phi]k^2r \\ & + i\omega\mu c_{v1}v[-B_v\sin v\phi + C_v\cos v\phi]k^v r^{v-1} \\ & + i\omega\mu c_{v+1,1}(v+1)[-B_{v+1}\sin(v+1)\phi \\ & + C_{v+1}\cos(v+1)\phi]k^{v+1}r^v + \cdots.\end{aligned} \tag{6.19b}$$

$$
\begin{aligned}
E_\phi = & -i\omega\mu c_{11}k[B_1\cos\phi + C_1\sin\phi] \\
& - 2i\omega\mu\{c_{02}B_0 + c_{21}[B_2\cos 2\phi + C_2\sin 2\phi]\}k^2 r \\
& - i\omega\mu c_{\nu 1}\nu[B_\nu\cos\nu\phi + C_\nu\sin\nu\phi]k^\nu r^{\nu-1} \\
& - i\omega\mu c_{\nu+1,1}(\nu+1)[B_{\nu+1}\cos(\nu+1)\phi \\
& + C_{\nu+1}\sin(\nu+1)\phi]k^{\nu+1}r^\nu \\
& - i\omega\mu(\nu+2)\{c_{\nu 2}[B_\nu\cos\nu\phi + C_\nu\sin\nu\phi] \\
& + c_{\nu+2,1}[B_{\nu+2}\cos(\nu+2)\phi + C_{\nu+2}\sin(\nu+2)\phi]\}k^{\nu+2}r^{\nu+1} + \ldots.
\end{aligned}
\tag{6.19c}
$$

Now let us insert these expressions of the field components into the first boundary condition (6.5a). According to the assumptions made in this subsection, on S we have $\phi = \theta$ and

$$\mathbf{n} = \mathbf{e}_\phi, \quad H_n = 0, \quad E_n = E_\phi, \quad \mathbf{E}_\mathrm{t} = E_r\mathbf{e}_r, \tag{6.20a}$$

$$\mathbf{n}\times\mathbf{H} = \mathrm{H_z}\mathbf{e}_r, \qquad \mathrm{grad}E_n = \partial E_\phi/\partial r\mathbf{e}_r. \tag{6.20b}$$

Here $\mathbf{e}_\phi, \mathbf{e}_r$, and $\mathbf{e}_z$ stand for the usual unit coordinate vectors. If one arranges the resulting expression of (6.5a) as a power series in r, then the coefficients of all powers becomes naught because the boundary condition is satisfied identically for all $r \in (0, r_0)$. Thus the coefficients of the powers $r^{\nu-2}, r^{\nu-1}, r^\nu$, and r^0 permit us to write

$$f(a_7, a_8) = 0, \quad g(a_3, a_4) + g_1(a_7, a_8) = 0, \tag{6.21a}$$

$$f(a_1', a_2') + f_1(a_3, a_4, a_7, a_8) = 0, \tag{6.21b}$$

and

$$h(a_1, a_2, a_3, a_4, a_7, a_8, \theta, \mu) = 0, \tag{6.22}$$

where we put

$$\mathrm{f}(a, a') = a[P\sin\nu\theta - Q\cos\nu\theta] + a'[P^*\sin\nu\theta - Q^*\cos\nu\theta] \tag{6.23a}$$

$$\mathrm{g}(a, a') = a[P\cos\nu\theta + Q\sin\nu\theta] + a'[P^*\cos\nu\theta + Q^*\sin\nu\theta] \tag{6.23b}$$

$$a_1' = k^2 a_1/\mu, \qquad a_2' = k^{*2}a_2/\mu* \tag{6.23c}$$

and

$$
\begin{aligned}
h(a_1, a_2, a_3, a_4, a_7, a_8, \theta, \mu) = & [a_1 + (i/2)\omega\mu a_7]k^2 B_0 \\
& + [a_2 + (i/2)\omega\mu^* a_8]k^{*2}B_0^* +
\end{aligned}
$$

$$+ a_3[P_1 \cos\theta + Q_1 \sin\theta] + a_4[P_1^* \cos\theta + Q_1^* \sin\theta]$$
$$+ a_7[Q_2 \cos 2\theta - P_2 \sin 2\theta] + a_8[Q_2^* \cos 2\theta - P_2^* \sin 2\theta] \quad (6.24)$$

with

$$P = i\omega\mu c_{\nu 1}\nu k^\nu C_\nu, \qquad Q = -i\omega\mu c_{\nu 1}\nu k^\nu B_\nu, \quad (6.25a)$$
$$P^* = i\omega\mu^* c_{\nu 1}\nu k^{*\nu} C_\nu^*, \qquad Q^* = -i\omega\mu^* c_{\nu 1}\nu k^{*\nu} B_\nu^*, \quad (6.25b)$$

and

$$P_1 = i\omega\mu c_{11} k C_1, \qquad Q_1 = -i\omega\mu k c_{11} B_1, \quad (6.26a)$$
$$P_1^* = i\omega\mu^* c_{11} k^* C_1^*, \qquad Q_1^* = -i\omega\mu^* c_{11} k^* B_1^*, \quad (6.26b)$$
$$P_2 = 2i\omega\mu c_{21} k^2 C_2, \qquad Q_2 = -2i\omega\mu c_{21} k^2 \mathrm{B}_2, \quad (6.27a)$$
$$P_2^* = 2i\omega\mu^* c_{21} k^{*2} C_2^*, \qquad Q_2^* = -2i\omega\mu^* c_{21} k^{*2} B_2^*. \quad (6.27b)$$

As to the functions $f_1(a_3, a_4, a_7, a_8)$ and $g_1(a_7, a_8)$, they are homogeneous linear functions of their arguments. In what follows, we will not need the explicit expressions of them (see problem 1 below).

If $|a_7| + |a_8| \neq 0$, then the first equation in (6.21) yields one of the equations that will permit us to determine the minimal exponent ν. In this case the other equations in (6.21), which involve more unknown coefficients, become useless. If, on the contrary, $a_7 = a_8 = 0$ but $|a_3| + |a_4| \neq 0$, then the above-mentioned equation is replaced by the second one, which is reduced to $g(a_3, a_4) = 0$. Finally, in the case when all the four coefficients a_7, a_8, a_3, and a_4 are zero, the third equation in (6.21) yields $f(a_1', a_2') = 0$. It is obvious that this is the last possibility because $a_1 = a_2 = 0$ is not possible since one has already supposed that $a_3 = a_4 = a_7 = a_8 = 0$.

All the results obtained above can be summarized as below:

$$\text{Cond.(6.5a) on } S \Rightarrow \text{if } |a_7| + |a_8| \neq 0 \Rightarrow f(a_7, a_8) = 0$$
$$\text{if } |a_7| + |a_8| = 0 \Rightarrow \text{when } |a_3| + |a_4| \neq 0 \Rightarrow g(a_3, a_4) = 0$$
$$\text{when } |a_3| + |a_4| = 0 \Rightarrow \mathrm{f}(a_1', a_2') = 0.$$

The discussion made for (6.5a) can also be repeated for (6.5b). The result is quite the same except that the constants $a_i (i = 1, 2, \ldots, 6)$ are now replaced by b_i.

Consider finally the conditions written on Σ. They give the similar results with two changes. First, $\phi = -\theta$ when one approaches Σ from the region B while $\phi = 2\pi - \theta$ when one approaches Σ from B^* (see

Fig. 6.2). Second, the normal on Σ is directed now into the region B. Thus the result can be shown schematically as follows:

$$\begin{aligned}&\text{Cond. (6.6a) on } \Sigma \Rightarrow \text{if } |\alpha_7| + |\alpha_8| \neq 0 \Rightarrow \psi(\alpha_8, \alpha_7) = 0\\ &\quad \text{if } |\alpha_7| + |\alpha_8| = 0 \Rightarrow \text{when } |\alpha_3| + |\alpha_4| \neq 0 \Rightarrow \chi(\alpha_4, \alpha_3) = 0\\ &\qquad\qquad\qquad\qquad\quad \text{when } |\alpha_3| + |\alpha_4| = 0 \Rightarrow \psi(\alpha_2', \alpha_1') = 0.\end{aligned}$$

Here we put

$$\begin{aligned}\psi(\alpha, \alpha') &= -\alpha[P \sin \nu\theta + Q \cos \nu\theta]\\ &\quad + \alpha'[P^* \sin \nu(2\pi - \theta) - Q^* \cos \nu(2\pi - \theta)], \qquad (6.28a)\\ \chi(\alpha, \alpha') &= \alpha[P \cos \nu\theta - Q \sin \nu\theta]\\ &\quad + \alpha'[P^* \cos \nu(2\pi - \theta) + Q^* \sin \nu(2\pi - \theta)] \qquad (6.28b)\end{aligned}$$

and

$$\alpha_1' = k^{*2}\alpha_1/(\omega\mu^*\nu), \qquad \alpha_2' = k^2\alpha_2/(\omega\mu\nu). \qquad (6.28c)$$

Note that the same results are also valid when $\alpha_i\,(i = 1, 2, \ldots, 6)$ are replaced by β_i.

PROBLEMS

1. Compute the coefficients of the terms of orders $r^{\nu-1}$ and r^{ν} in the boundary condition (6.5a) and show that

$$\begin{aligned}(i\omega\nu)^{-1}g_1(a_7, a_8) &= a_7\{-\mu c_{\nu+1,1}(\nu + 1)[B_{\nu+1} \cos(\nu + 1)\phi\\ &\qquad + C_{\nu+1} \sin(\nu + 1)\phi]k^{\nu+1}\}\\ &\quad + a_8\{-\mu^* c_{\nu+1,1}(\nu + 1)[B^*{}_{\nu+1} \cos(\nu + 1)\phi\\ &\qquad + C^*{}_{\nu+1} \sin(\nu + 1)\phi]k^{*\nu+1}\}\end{aligned}$$

and

$$\begin{aligned}(i\omega\nu)^{-1}f_1(a_3, a_4, a_7, a_8) &= a_3\{i\omega\mu c_{\nu+1,1}(\nu + 1)[-B_{\nu+1} \sin(\nu + 1)\phi\\ &\qquad + C_{\nu+1} \cos(\nu + 1)\phi]k^{\nu+1}\}\\ &\quad + a_4\{i\omega\mu^* c_{\nu+1,1}(\nu + 1)[-B^*{}_{\nu+1} \sin(\nu + 1)\phi\\ &\qquad + C^*{}_{\nu+1} \cos(\nu + 1)\phi]k^{*\nu+1}\}\\ &\quad + a_7\{-i\omega\mu(\nu + 2)(\nu + 1)\{c_{\nu 2}[B_\nu \cos \nu\phi\\ &\qquad + C_\nu \sin \nu\phi] +\end{aligned}$$

$$
\begin{aligned}
&+ c_{\nu+2,1}[B_{\nu+2}\cos(\nu+2)\phi \\
&+ C_{\nu+2}\sin(\nu+2)\phi]\}k^{\nu+2}\} \\
&+ a_8\{-i\omega\mu^*(\nu+2)(\nu+1)\{c_{\nu 2}[B^*{}_{\nu}\cos\nu\phi \\
&+ C^*{}_{\nu}\sin\nu\phi] \\
&+ c_{\nu+2,1}[B^*{}_{\nu+2}\cos(\nu+2)\phi \\
&+ C^*{}_{\nu+2}\sin(\nu+2)\phi]\}k^{*\nu+2}\}.
\end{aligned}
$$

2. Compute the coefficients of the terms of orders r^0 in the boundary condition (6.5a) and find (6.22) with (6.24). Note that $c_{01} = 1$ and $c_{02} = -1/4$.

D. An Example

In order to clarify the above discussion, let us reconsider a generalization of a configuration studied first by Lang [53] and Braver et al. [57] for some particular cases and then in Idemen [59] for a more general case. In this example, S is perfectly conducting while Σ consists of a resistive half-plane. One supposes also that the spaces B and B^* have the same constitutive parameters. Then, the boundary condition tables are

$$
S \Rightarrow \begin{bmatrix} 0 & 0 & 1 & 0 & 0 & 0 & 0 & 0 \\ 0 & 0 & 0 & 1 & 0 & 0 & 0 & 0 \end{bmatrix}, \quad \Sigma \Rightarrow \begin{bmatrix} 0 & 0 & 1 & -1 & 0 & 0 & 0 & 0 \\ 1 & -1 & 1/R & 0 & 0 & 0 & 0 & 0 \end{bmatrix},
$$

which explicitly yield (see (4.19a))

$$
\mathbf{E}_t(r,\theta) = \mathbf{E}_t^*(r,\theta) = 0 \quad \text{on } S,
$$

and

$$
\begin{aligned}
\mathbf{E}_t(\mathbf{r},-\theta) &= \mathbf{E}_t^*(r,2\pi-\theta) \\
&= -R\mathbf{n}\times[\mathbf{H}(r,-\theta)-\mathbf{H}^*(r,2\pi-\theta)] \quad \text{on } \Sigma.
\end{aligned}
$$

From these tables, one concludes that on S one has $a_7 = a_8 = 0$ and $b_7 = b_8 = 0$ while $a_3 = 1, a_4 = 0$, and $b_3 = 0, b_4 = 1$, which permit us to write $g(1,0) = 0$ and $g(0,1) = 0$, respectively. Similarly, on Σ one has $\alpha_7 = \alpha_8 = 0$, $\alpha_3 = 1$, $\alpha_4 = -1$ and $\beta_7 = \beta_8 = 0$, $\beta_3 = 1/R, \beta_4 = 0$, which yield $\chi(-1,1) = 0$ and $\chi(0,1/R) = 0$. The equations $g(1,0) = 0, g(0,1) = 0, \chi(-1,1) = 0$, and $\chi(0,1/R) = 0$ are written explicitly as follows:

$$
\begin{aligned}
&P\cos\nu\theta + Q\sin\nu\theta = 0, \\
&P^*\cos\nu\theta + Q^*\sin\nu\theta = 0, \\
&-P\cos\nu\theta + Q\sin\nu\theta + P^*\cos\nu(2\pi-\theta) + Q^*\sin\nu(2\pi-\theta) = 0,
\end{aligned}
$$

$$P^* \cos \nu(2\pi - \theta) + Q^* \sin \nu(2\pi - \theta) = 0.$$

The compatibility of these homogeneous equations requires

$$\Delta(\nu) = \begin{vmatrix} \cos \nu\theta & \sin \nu\theta & 0 & 0 \\ 0 & 0 & \cos \nu\theta & \sin \nu\theta \\ -\cos \nu\theta & \sin \nu\theta & \cos \nu(2\pi - \theta) & \sin \nu(2\pi - \theta) \\ 0 & 0 & \cos \nu(2\pi - \theta) & \sin \nu(2\pi - \theta) \end{vmatrix},$$

which is merely equal to

$$\Delta(\nu) = -\sin 2\nu\theta \sin 2\nu(\pi - \theta) = 0$$

and gives

$$\nu = \min\{\pi/[2\theta], \pi/[2(\pi - \theta)]\}$$

whenever $\theta \neq \pi/2$ (the case of $\theta = \pi/2$ will be discussed at the end of Section 6.3.1.G below).

E. Determination of the Minimal Exponent and Canonical Types

The four equations such as $f = 0, g = 0$, and so on, derived in Section 6.3.1.C above constitute a system of linear and homogeneous algebraic equations in terms of the constants P, Q, P^*, and Q^*. Since at least one of these latter coefficients is different from zero, the determinant formed of the coefficients appearing in these equations, say $\Delta(\nu)$, is zero. It is the equation $\Delta(\nu) = 0$ that will permit us to determine the minimal exponent ν. Before going further into the details of computation, we want to examine first the number of the possible cases.

By considering the schemas mentioned in Section 6.3.1.*C* above, we can claim that the condition (6.5a) written for S can exhibit three different cases. The same is also true for the other boundary conditions written on S and Σ. Thus the total number of the physically different cases is $3^4 = 81$. Fortunately, all these physically different cases are not mathematically different because the left-hand sides of the equations are always f or g and ψ or χ and the numerical values of the constants a_i, b_i, α_i, and β_i taking place in these equations are not mathematically important. If we signify each case by indicating the left-hand sides of the four equations related to this case, then we can distinguish the following 16 different cases:

$$\begin{array}{llll} (ff\psi\psi) & (ff\psi\chi) & (ff\chi\psi) & (ff\chi\chi), \\ (fg\psi\psi) & (fg\psi\chi) & (fg\chi\psi) & (fg\chi\chi), \\ (gf\psi\psi) & (gf\psi\chi) & (gf\chi\psi) & (gf\chi\chi), \\ (gg\psi\psi) & (gg\psi\chi) & (gg\chi\psi) & (gg\chi\chi). \end{array}$$

It is evident that these apparently different cases will not also lead to mathematically different results because an interchange between the rows in a determinant does not cause to any change in the value of the determinant. Thus the cases that have to be considered in full detail are only the following nine *canonical* types:

$$\begin{array}{lll} (ff\psi\psi) & (ff\psi\chi) & (ff\chi\chi), \\ (fg\psi\psi) & (fg\psi\chi) & (fg\chi\chi), \\ (gg\psi\psi) & (gg\psi\chi) & (gg\chi\chi). \end{array}$$

The determinants corresponding to all these canonical cases except the fifth one can be computed very easily. The results will be recapitulated in Table 6.1. From this table it is evidently seen that in four cases (i.e., the types $(ff\psi\psi), (gg\chi\chi), (ff\chi\chi)$, and $(gg\psi\psi)$) the value of v is determined only via the geometry of the wedge (i.e., via the angle θ), while in the other cases the constitutive parameters of the boundary of the wedge as well as those of the regions B and B^* affect also the value of v. To find the numerical value of v in these rather complicated cases except the case $(fg\psi\chi)$, one has to solve a trigonometric equation of the form

$$\sin 2v(\pi-\theta)\cos 2v\theta = A\sin 2v\theta\cos 2v(\pi-\theta),$$

which can be solved, in general, through numerical methods. As to the most complicated case $(fg\psi\chi)$, the resulting equation is as follows:

$$\Delta(v)=\begin{vmatrix} a\sin v\theta & -a\cos v\theta & a'\sin v\theta & -a'\cos v\theta \\ b\cos v\theta & b\sin v\theta & b'\cos v\theta & b'\sin v\theta \\ \alpha'\sin v\theta & \alpha'\cos v\theta & -\alpha\sin v(2\pi-\theta) & \alpha\cos v(2\pi-\theta) \\ \beta'\cos v\theta & -\beta'\sin v\theta & \beta\cos v(2\pi-\theta) & \beta\sin v(2\pi-\theta) \end{vmatrix} = 0. \tag{6.29}$$

Note that the constants appearing in (6.29) are the parameters that take place in the expressions of the equations $(fg\psi\chi)$ successively, namely,

$$f(a,a')=0,\quad g(b,b')=0,\quad \psi(\alpha',\alpha)=0,\quad \chi(\beta',\beta)=0.$$

Table 6.1.
Minimal Exponent ν for the Canonical Types of a Penetrable Wedge

Type	Typical Equations	ν	Validity Conditions
$(ff\psi\psi)$	$f(a,a')=f(b,b')=0$ $\Psi(\alpha',\alpha)=\Psi(\beta',\beta)=0$	$\min\{\pi/(2\theta), \pi/(2(\pi-\theta))\}$ $(1/2<\nu<1)$ $\nu=1$ (double $\Rightarrow$ log. singularity)	$ab'-a'b\neq 0, \theta\neq\pi/2$ $\alpha\beta'-\alpha'\beta\neq 0$ $\theta\to\pi/2$
$(gg\chi\chi)$	$g(a,a')=g(b,b')=0$ $\chi(\alpha',\alpha)=\chi(\beta',\beta)=0$	$\min\{\pi/(2\theta), \pi/(2(\pi-\theta))\}$ $(1/2<\nu<1)$ $\nu=1$ (double $\Rightarrow$ log. singularity)	$ab'-a'b\neq 0, \theta\neq\pi/2$ $\alpha\beta'-\alpha'\beta\neq 0$ $\theta\to\pi/2$
$(ff\chi\chi)$	$f(a,a')=f(b,b')=0$ $\chi(\alpha',\alpha)=\chi(\beta',\beta)=0$	$\min\{\pi/(4\theta), \pi/(4(\pi-\theta))\}$ $(1/4<\nu<1/2)$	$ab'-a'b\neq 0$ $\alpha\beta'-\alpha'\beta\neq 0$
$(gg\psi\psi)$	$g(a,a')=g(b,b')=0$ $\psi(\alpha',\alpha)=\psi(\beta',\beta)=0$	$\min\{\pi/(4\theta), \pi/(4(\pi-\theta))\}$ $(1/4<\nu<1/2)$	$ab'-a'b\neq 0$ $\alpha\beta'-\alpha'\beta\neq 0$
$(ff\psi\chi)$	$f(a,a')=f(b,b')=0$ $\psi(\alpha',\alpha)=\chi(\beta',\beta)=0$	$tg2\nu(\pi-\theta)=Atg2\nu\theta$ $A=-\alpha'\beta/(\alpha\beta')$	$ab'-a'b\neq 0$
$(fg\psi\psi)$	$f(a,a')=g(b,b')=0$ $\psi(\alpha',\alpha)=\psi(\beta',\beta)=0$	$tg2\nu(\pi-\theta)=Atg2\nu\theta$ $A=-ab'/(a'b)$	$\alpha\beta'-\alpha'\beta\neq 0$
$(fg\chi\chi)$	$f(a,a')=g(b,b')=0$ $\chi(\alpha',\alpha)=\chi(\beta',\beta)=0$	$tg2\nu(\pi-\theta)=Atg2\nu\theta$ $A=-a'b/(ab')$	$\alpha\beta'-\alpha'\beta\neq 0$
$(gg\psi\chi)$	$g(a,a')=g(b,b')=0$ $\psi(\alpha',\alpha)=\chi(\beta',\beta)=0$	$tg2\nu(\pi-\theta)=Atg2\nu\theta$ $A=-\alpha\beta'/(\alpha'\beta)$	$ab'-a'b\neq 0$
$(fg\psi\chi)$	$f(a,a')=g(b,b')=0$ $\psi(\alpha',\alpha)=\chi(\beta',\beta)=0$	Equation (6.29)	

Remark. The above analyses were made under the assumption (6.15b). If the minimal root of $\Delta(\nu) = 0$ consists of an integer $\nu_n = n$, then we can suppose the present situation as the limiting case of a certain configuration for which $\nu \to \nu_n$. Thus all the results to be obtained below with the restriction (6.15b) become valid for all minimal ν *in this sense*. For example, when $\theta \to \pi/2$, the minimal exponent found in the example considered in Section 6.3.1.D above tends to 1.

F. Determination of the Constants B_0 and B_0^*

From (6.16c) it is obvious that when $B_0 \neq 0$, the dominant behavior of E_z is not like $O(r^\nu)$ but instead like $O(1)$. Therefore one has always to check whether B_0 and B_0^* are zero or not. The first relations involving these constants is given by (6.24). To that equation are added three equations obtained by replacing a_j by b_j, α_j, and β_j. The relations connected with the boundary Σ require also the change $\theta \to -\theta$. Thus one has

$$h(a_1, a_2, a_3, a_4, a_7, a_8, \theta, \mu) = 0, \tag{6.30a}$$

$$h(b_1, b_2, b_3, b_4, b_7, b_8, \theta, \mu) = 0, \tag{6.30a'}$$

and

$$h(\alpha_2, \alpha_1, \alpha_4, \alpha_3, \alpha_8, \alpha_7, -\theta, \mu) = 0, \tag{6.30b}$$

$$\mathrm{h}(\beta_2, \beta_1, \beta_4, \beta_3, \beta_8, \beta_7, -\theta, \mu) = 0 \tag{6.30b'}$$

with $h(a_1, a_2, a_3, a_4, a_7, a_8, \theta, \mu)$ given by (6.24). The constants (P_1, Q_1, P_1^*, Q_1^*) and (P_2, Q_2, P_2^*, Q_2^*) are the coefficients in the terms corresponding to the spectral points $\nu_1 = 1$ and $\nu_2 = 2$, respectively. If the points ν_1 and ν_2 belong to the spectrum, then the four equations (6.30a–b′) involve two or six or ten unknown constants, depending on the values of the coefficients appearing in the boundary conditions (6.5a)–(6.6b). From this system one gets, in general, $B_0 \neq 0$ and $B_0^* \neq 0$. But, in some cases they give $B_0 = B_0^* = 0$. The exact values of these constants will be fixed, of course, by the source of the field.

As an example reconsider the configuration studied in Section 6.3.1.D above. If one inserts the constants a_j, b_j, α_j, and β_j, given in the boundary conditions tables connected with the surfaces S and Σ

into (6.30a,b), then one gets (note that in this example $k = k^*$)

$$\begin{aligned}
&P_1 \cos\theta + Q_1 \sin\theta = 0,\\
&P_1^* \cos\theta + Q_1^* \sin\theta = 0,\\
&[P_1 \cos\theta - Q_1 \sin\theta] - [P_1^* \cos\theta - Q_1^* \sin\theta] = 0,\\
&k^2 B_0 - k^2 B_0^* - \frac{1}{R}[P_1^* \cos\theta - Q_1^* \sin\theta] = 0.
\end{aligned}$$

From these equations one concludes that B_0 and B_0^* may be different from zero. In other words, $H_z = O(1)$. In the case of $\theta = \pi/2$, one gets $Q_1 = Q_1^* = 0$ and $B_0 = B_0^*$.

G. Confluence of the Zeros of $\Delta(v)$. Logarithmic Singularities

The minimal exponent ν that satisfies a certain equation $\Delta(\nu) = 0$, where $\Delta(\nu)$ denotes an entire function of ν, is obviously a function of the parameters such as $\theta, a_i, \alpha_i, \varepsilon', \varepsilon'^*$, and so on. Let τ denote any one of these parameters. If we write now the above-mentioned entire function $\Delta(\nu)$ more explicitly as $\Delta(\nu, \tau)$ and consider all the roots of the equation $\Delta(\nu, \tau) = 0$, which are in general of infinite number, then we obtain a relation similar to what is shown in Fig. 6.3. The minimal exponent ν is, in general, such that the point (τ, ν) is an ordinary point of the curve which represents all branches of the function $\nu = \nu(\tau)$. In this case $\nu = \nu(\tau)$ generates the solutions (6.16a–c) and (6.17a–c). When τ tends, however, to a certain critical value, say τ_0, the point (τ_0, ν_0) may become a branch point of the curve mentioned above, such that at this point

$$\nu' \to \nu'' \to \nu_0 \qquad \text{when } \tau \to \tau_0. \tag{6.31}$$

Here ν' and ν'' stand for the first two solutions of the equation $\Delta(\nu, \tau) = 0$, that is,

$$\Delta(\nu', \tau) = 0, \qquad \Delta(\nu'', \tau) = 0. \tag{6.32}$$

Now it is an easy matter to show that for the couple (τ_0, ν_0) one has

$$\partial/\partial\nu\, \Delta(\nu, \tau_0) = 0, \qquad \nu = \nu_0, \tag{6.33}$$

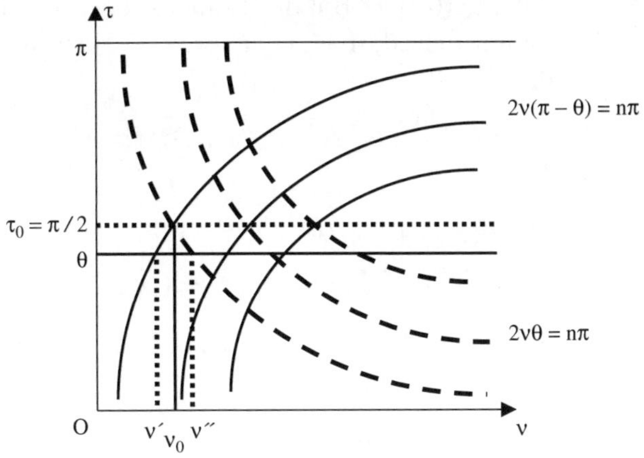

Figure 6.3. Variation of the roots of $\Delta(\nu, \tau) = 0$ with $\tau = \theta$ for the configuration considered in Sec 6.3.1.D. (---) shows the hyperbolas $2\nu\theta = n\pi$ while (——) are the hyperbolas $2\nu(\pi - \theta) = n\pi (n = 1, 2, \ldots)$.

which implies that ν_0 is a *double* root of the equation $\Delta(\nu, \tau_0) = 0$. Indeed, from (6.32) we write first

$$\begin{aligned} 0 &= [\Delta(\nu', \tau) - \Delta(\nu'', \tau)]/(\nu' - \nu'') \\ &= [\partial/\partial\nu'' \Delta(\nu'', \tau)(\nu' - \nu'') + o((\nu' - \nu''))]/(\nu' - \nu''), \end{aligned}$$

the continuity of $\Delta(\nu, \tau)$ being taken into account. If one makes now $\tau \to \tau_0$, because of (6.31), one gets

$$\partial/\partial\nu'' \Delta(\nu'', \tau_0) = 0, \qquad \nu'' = \nu_0,$$

which is nothing but (6.33).

Now return to the field expressions given by (6.16a–c) and suppose that $\tau = \tau_0$. Since this *critical case* can be thought as being the limiting case when $\tau \to \tau_0$, we have to consider both exponents $\nu' \to \nu_0$ and $\nu'' \to \nu_0$ together and write

$$\begin{aligned} E_r \sim \lim_{\tau \to \tau_0} \{ & r^{\nu' - 1}[P' \cos \nu'\phi + Q' \sin \nu'\phi] \\ & + r^{\nu'' - 1}[P'' \cos \nu''\phi + Q'' \sin \nu''\phi]\}. \end{aligned} \tag{6.34}$$

Here P', Q', P'', and Q'' denote certain constants that can also depend on the parameter τ or, inversely, on ν' and ν''. If these coefficients individually tend to certain finite limits, then (6.34) gives

$$E_r \sim r^{\nu_0 - 1}[P \cos \nu_0\phi + Q \sin \nu_0\phi], \tag{6.35}$$

which is quite similar to (6.16a). But this is not the only possibility that produces a finite limit. Indeed, if P', Q', P'', and Q'' tend individually to infinity such that

$$\begin{aligned}
P'(v') &\to P_0/(v' - v'') + P_1 \\
P''(v'') &\to P_0/(v'' - v') + P_2 \\
Q'(v') &\to Q_0/(v' - v'') + Q_1 \\
Q''(v'') &\to Q_0/(v'' - v') + Q_2,
\end{aligned}$$

where P_0, Q_0, P_1, Q_1, P_2, and Q_2 are certain constants, then (6.34) will also produce a finite limit, namely,

$$\begin{aligned}
E_r &\sim r^{v_0 - 1}[P \cos v_0\phi + Q \sin v_0\phi] \\
&\quad + P_0 \partial/\partial v'[r^{v'-1} \cos v'\phi] + Q_0 \partial/\partial v'[r^{v'-1} \sin v'\phi],\ v' = v_0 \\
&\sim \mathrm{r}^{v_0 - 1}[P \cos v_0\phi + Q \sin v_0\phi] \\
&\quad + r^{v_0 - 1}[Q_0\phi \cos v_0\phi - P_0\phi \sin v_0\phi] \\
&\quad + r^{v_0 - 1} \ln r[P_0 \cos v_0\phi + Q_0 \sin v_0\phi]. \qquad (6.36)
\end{aligned}$$

The first term in (6.36) is quite similar to that in (6.16a). Therefore, by substituting v_0 again with v we can claim that when the minimal exponent v is a double root of the equation $\Delta(v) = 0$, the expressions (6.16a–c) are replaced by the following *confluent* expressions:

$$\begin{aligned}
E_r &\sim r^{v-1}[P \cos v\phi + Q \sin v\phi] \\
&\quad + r^{v-1}[-P_0\phi \sin v\phi + Q_0\phi \cos v\phi] \\
&\quad + r^{v-1} \ln r[P_0 \cos v\phi + Q_0 \sin v\phi] \qquad (6.37\text{a}) \\
E_\phi &\sim r^{v-1}[Q \cos v\phi - P \sin v\phi] \\
&\quad - r^{v-1}[Q_0\phi \sin v\phi + P_0\phi \cos v\phi] \\
&\quad + r^{v-1} \ln r[Q_0 \cos v\phi - P_0 \sin v\phi] \qquad (6.37\text{b}) \\
H_z &\sim k^2 B_0 + ik^2/(v\omega\mu) r^v [Q \cos v\phi - P \sin v\phi] \\
&\quad - ik^2/(v\omega\mu) r^v [Q_0\phi \sin v\phi + P_0\phi \cos v\phi] \\
&\quad + ik^2/(v\omega\mu) r^v \ln r[Q_0 \cos v\phi - P_0 \sin v\phi]. \qquad (6.37\text{c})
\end{aligned}$$

Note that these expressions involve also such terms as $r^{v-1}[\ln r \cos v\phi - \phi \sin v\phi]$ and $r^{v-1}[\ln r \sin v\phi + \phi \cos v\phi]$, which are not in separated form (cf. (6.8a,b)).

The relations valid in the region B^* (i.e., for $\phi \in (\theta, 2\pi - \theta)$), which corresponds to (6.17a–c), are obtained from (6.37a–c) through the substitution

$$k \to k^*, \quad B_0 \to B_0^*, \quad \mu \to \mu^*, \quad P \to P^*, \tag{6.38a}$$

$$Q \to Q^*, \quad P_0 \to P_0^*, \quad Q_0 \to Q_0^*. \tag{6.38b}$$

Example As an example, reconsider the case studied in Section 6.3.1.D above. The first two roots of the equation $\Delta(\nu) = 0$ are $\nu' = \pi/[2\theta]$ and $\nu'' = \pi/[2(\pi - \theta)]$. They obviously depend on the angle θ and when $\theta \to \pi/2, \nu' \to \nu'' \to 1$. That means that in the case of a two-part plane with the above-mentioned physical properties there is a *coalescence* of the roots, which gives rise to *logarithmic singularities*. Beyond this critical case, the minimal exponent is always simple and results *algebraic singularities*. It is worthwhile to remark that this configuration was also considered by Lang [53], Braver et al. [57], and Idemen [59]. Lang has considered the cases of $\theta = 3\pi/4$ (or $\theta = \pi/4$) and $\theta = \pi/2$, while Braver et al. have considered only the case $\theta = \pi/2$. Our results are quite identical to those obtained by Braver et al. for $\theta = \pi/2$. As to the results given by Lang for $\theta = 3\pi/4$ and $\theta = \pi/2$, they all are quite different from our results.

6.3.2 The *E* Case

Consider now the case when the electric field is parallel to the edge. In this case the nonzero components of the electromagnetic field (i.e., E_z, H_r, and H_ϕ) can be expressed in terms of a potential function u satisfying (6.7a) as follows [63]:

$$E_z = k^2 u, \qquad H_r = -i\omega\varepsilon'/r\partial u/\partial\phi, \qquad H_\phi = i\omega\varepsilon'\partial u/\partial r. \tag{6.39a}$$

Here $\varepsilon' = (\varepsilon + i\sigma/\omega)$ denotes the complex permittivity of the region B. By replacing k and ε' in (6.39a) by k^* and ε'^*, respectively, one gets the expressions valid in B^*. It is evident that these expressions can be obtained from (6.11) by making the substitutions

$$-\mu \to \varepsilon', \qquad H_z \to E_z, \qquad E_r \to H_r, \qquad E_\phi \to H_\phi. \tag{6.39b}$$

Therefore, by making also these substitutions in the results obtained in Section 6.3.1, we get the results that will be valid in this case. When there is no confluence—that is, when ν is a simple root of

$\Delta(\nu) = 0$—we have in B, $\phi \in (-\theta, \theta)$ and

$$H_r \sim r^{\nu-1}[P \cos \nu\phi + Q \sin \nu\phi], \tag{6.40a}$$

$$H_\phi \sim r^{\nu-1}[Q \cos \nu\phi - P \sin \nu\phi], \tag{6.40b}$$

and

$$E_z \sim k^2 B_0 - ik^2/(\nu\omega\varepsilon')r^\nu[Q \cos \nu\phi - P \sin \nu\phi]. \tag{6.40c}$$

The expressions valid in B^* are obtained from the above ones by replacing P, Q, k, B_0, and ε' by P^*, Q^*, k^*, B_0^*, and $\varepsilon^{*\prime}$, respectively.

In the case of confluence—that is, when ν is a double root of $\Delta(\nu) = 0$—(6.40a–c) are replaced by

$$H_r \sim r^{\nu-1}[P \cos \nu\phi + Q \sin \nu\phi] + r^{\nu-1}[-P_0\phi \sin \nu\phi + Q_0\phi \cos \nu\phi] + r^{\nu-1} \ln r[P_0 \cos \nu\phi + Q_0 \sin \nu\phi], \tag{6.41a}$$

$$H_\phi \sim r^{\nu-1}[Q \cos \nu\phi - P \sin \nu\phi] - r^{\nu-1}[Q_0\phi \sin \nu\phi + P_0\phi \cos \nu\phi] + r^{\nu-1} \ln r[Q_0 \cos \nu\phi - P_0 \sin \nu\phi], \tag{6.41b}$$

$$E_z \sim k^2 B_0 - ik^2/(\nu\omega\varepsilon')r^\nu[Q \cos \nu\phi - P \sin \nu\phi] + ik^2/(\nu\omega\varepsilon')r^\nu[Q_0\phi \sin \nu\phi + P_0\phi \cos \nu\phi] - ik^2/(\nu\omega\varepsilon')r^\nu \ln r[Q_0 \cos \nu\phi - P_0 \sin \nu\phi]. \tag{6.41c}$$

The expressions valid in the region B^* are obtained from (6.41a–c) by considering $\phi \in (\theta, 2\pi - \theta)$ and making the substitutions

$$k \to k^*, \qquad \mu \to \mu*, \qquad P \to P^*, \tag{6.42a}$$

$$Q \to Q^*, \qquad P_0 \to P^*{}_0, \qquad Q_0 \to Q^*{}_0. \tag{6.42b}$$

As to the boundary conditions on S and Σ, we have now

$$\mathbf{n} = \mathbf{e}_\phi, \qquad H_n = H_\phi, \qquad \mathbf{n} \times \mathbf{H} = -H_r\mathbf{e}_z, \qquad \mathbf{E_t} = E_z\mathbf{e}_z$$

$$E_n = 0, \qquad \text{grad}H_n = \partial/\partial \text{r} H_\phi \mathbf{e}_r, \qquad \mathbf{n} \times \text{grad}H_n = -\partial/\partial \text{r} H_\phi \mathbf{e}_z.$$

If we consider these relations in (6.5a) and (6.6b), and follow the same procedure as in Section 6.3.1.C above, then we get the following schemas:

Cond. (6.5a) on $S \Rightarrow$ if $|a_5| + |a_6| \neq 0 \Rightarrow f(a_5, a_6) = 0$

if $|a_5| + |a_6| = 0 \Rightarrow$ when $|a_1| + |a_2| \neq 0 \Rightarrow \text{g}(a_1, a_2) = 0$

when $|a_1| + |a_2| = 0 \Rightarrow \text{f}(a_3', a_4') = 0.$

$$\text{Cond. (6.6a) on } \Sigma \Rightarrow \text{ if } |\alpha_5| + |\alpha_6| \neq 0 \Rightarrow \psi(\alpha_6, \alpha_5) = 0$$
$$\text{if } |\alpha_5| + |\alpha_6| = 0 \Rightarrow \text{ when } |\alpha_1| + |\alpha_2| \neq 0 \Rightarrow \chi(\alpha_2, \alpha_1) = 0$$
$$\text{when } |\alpha_1| + |\alpha_2| = 0 \Rightarrow \psi(\alpha_4', \alpha_3') = 0.$$

Here we put

$$a_3' = k^2 a_3/(\nu\omega\varepsilon'), \qquad a_4' = k^{*2} a_4/(\nu\omega\varepsilon^{*\prime}) \tag{6.43a}$$
$$\alpha_3' = k^{*2} \alpha_3/(\nu\omega\varepsilon^{*\prime}), \qquad \alpha_4' = k^2 \alpha_4/(\nu\omega\varepsilon'). \tag{6.43b}$$

By making the substitutions $a_i \to b_i$ and $\alpha_i \to \beta_i$ in the above schemas, we obtain the results connected with the conditions (6.5b) and (6.6b). Therefore the discussion made in Section 6.3.1.D above as well as in the Table 6.1 are valid for this case also.

As to the equations satisfied by the constants B_0 and B_0^* (i.e., (6.30a,b)), they are replaced now by the following ones:

$$h(a_3, a_4, a_1, a_2, a_5, a_6, \theta, \varepsilon') = 0, \tag{6.44a}$$
$$h(b_3, b_4, b_1, b_2, b_5, b_6, \theta, \varepsilon') = 0, \tag{6.44a'}$$
$$h(\alpha_4, \alpha_3, \alpha_2, \alpha_1, \alpha_6, \alpha_5 - \theta, \varepsilon') = 0, \tag{6.44b}$$
$$h(\beta_4, \beta_3, \beta_2, \beta_1, \beta_6, \beta_5, -\theta, \varepsilon') = 0, \tag{6.44b'}$$

where $h(x, y, \ldots)$ is given by (6.24).

Example As a simple example consider again the configuration discussed in Section 6.3.1.D above. From the boundary conditions table given there it is obvious that

$$a_5 = a_6 = a_1 = a_2 = 0, \qquad a_3{}' = ik^2/(\nu\omega\varepsilon'), \qquad a_4{}' = 0$$
$$b_5 = b_6 = b_1 = b_2 = 0, \qquad b_3{}' = 0, \qquad b_4' = ik^2/(\nu\omega\varepsilon')$$

and

$$\alpha_5 = \alpha_6 = \alpha_1 = \alpha_2 = 0, \qquad \alpha_3' = -\alpha_4' = ik^2/(\nu\omega\varepsilon')$$
$$\beta_5 = \beta_6 = 0, \qquad \beta_1 = -\beta_2 = 1.$$

Hence the equations from which one finds the minimal exponent ν are as follows:

$$f(a_3', 0) = 0, \quad f(0, b_4') = 0, \quad \psi(\alpha_4', \alpha_3') = 0, \quad \chi(-1, 1) = 0.$$

This is obviously the $(ff\psi\chi)$ case indicated on the fifth row of Table 6.1. Since $A = -1$, the equation from which one finds ν reduces to $\sin 2\nu\pi = 0$, whence one gets $\nu = 1/2$. It is interesting to observe that in this case the minimal exponent ν does not depend on the angle θ whereas in the H case it was.

As to the constants B_0 and B_0^*, from (6.44a), one gets directly $B_0 = B_0^* = 0$.

6.4 THE WEDGE WITH IMPENETRABLE BOUNDARIES

If the boundaries of the wedge are impenetrable material sheets, then the electromagnetic fields inside B and B^* can exhibit different asymptotic behaviors when $r \to 0$. That means that the expressions in (6.16a–c) and (6.17a–c) are written with different exponents ν and ν^* inside B and B^*, respectively. Since the expressions valid for B^* can be obtained from those for B by making $k \to k^*$, in what follows we will consider only the region B.

Without loss of generality we can assume that

$$a_2 = a_4 = a_6 = a_8 = 0, \qquad b_i = 0 \qquad (i = 1, \ldots, 8), \tag{6.45a}$$

$$\alpha_1 = \alpha_3 = \alpha_5 = \alpha_7 = 0, \qquad \beta_i = 0 \qquad (i = 1, \ldots 8). \tag{6.45b}$$

In such a case, two of the four equations derived in Section 6.3 (i.e., those involving b_i and β_i) become identity while the remaining ones are reduced to the following equations:

In H-case $\Rightarrow$

$$f(a_7, 0) = 0 \quad \text{or} \quad g(a_3, 0) = 0 \quad \text{or} \quad f(a_1', 0) = 0,$$

$$\psi(\alpha_8, 0) = 0 \quad \text{or} \quad \chi(\alpha_4, 0) = 0 \quad \text{or} \quad \psi(\alpha_2', 0) = 0.$$

In E-case $\Rightarrow$

$$f(a_5, 0) = 0 \quad \text{or} \quad g(a_1, 0) = 0 \quad \text{or} \quad f(a_3', 0) = 0,$$

$$\psi(\alpha_6, 0) = 0 \quad \text{or} \quad \chi(\alpha_2, 0) = 0 \quad \text{or} \quad \psi(\alpha_4', 0) = 0.$$

This shows that the canonical types consist now only of $(f\psi), (f\chi)$, $(g\psi)$, and $(g\chi)$. The value of the minimal exponent ν corresponding to these cases are shown in Table 6.2. It is interesting to observe that the value of ν is always determined only by the geometry of the wedge (i.e., by the angle θ). Note that when the validity conditions indicated in Table 6.2 are met, the given value of ν is always in the interval (0,1). Otherwise, it becomes greater than $\nu_1 = 1$, which shows that the minimal exponent can be equal to $\nu_1 = 1$.

As to the equations satisfied by the constant B, it can be obtained from (6.30a,b) and (6.44a,b) by inserting (6.45a,b).

Table 6.2.
Minimal Exponent ν for the Canonical Types of an Impenetrable Wedge

Type	ν	Validity Conditions
$(f\psi), (g\chi)$	$\pi/(2\theta)$	$\theta > \pi/2$
$(f\chi), (g\psi)$	$\pi/(4\theta)$	$\theta > \pi/4$

6.5 EXAMPLES. APPLICATION TO HALF-PLANES

In the most important actual applications the boundaries consist of simple interfaces or material sheets with impedance, resistive, modified conducting, or magnetically conducting properties. The boundary condition tables (i.e., the coefficients a_i and b_i or α_i and β_i) related to these sheets are as follows:

$$\text{Simple interface}: \quad \begin{bmatrix} 0 & 0 & 1 & -1 & 0 & 0 & 0 & 0 \\ 1 & -1 & 0 & 0 & 0 & 0 & 0 & 0 \end{bmatrix}$$

$$\text{Impedance sheet}: \quad \begin{bmatrix} Z_1 & 0 & 1 & 0 & 0 & 0 & 0 & 0 \\ 0 & Z_2 & 0 & -1 & 0 & 0 & 0 & 0 \end{bmatrix}$$

$$\text{Resistive sheet [23]}: \quad \begin{bmatrix} 0 & 0 & 1 & -1 & 0 & 0 & 0 & 0 \\ 1 & -1 & 1/R & 0 & 0 & 0 & 0 & 0 \end{bmatrix}$$

Modified conducting sheet [64] :

$$\begin{bmatrix} 1 & -1 & 1/(2R) & 1/(2R) & 0 & 0 & 0 & 0 \\ 0 & 0 & 1 & -1 & 0 & 0 & A & A \end{bmatrix}$$

Magnetically conducting sheet [23] :

$$\begin{bmatrix} 1 & -1 & 0 & 0 & 0 & 0 & 0 & 0 \\ 1/G & 0 & -1 & 1 & 0 & 0 & 0 & 0 \end{bmatrix}$$

From these tables it is obviously seen that only the impedance boundaries are impenetrable. By using the values given in these tables in the results obtained in Sections 6.3 and 6.4, one gets the edge conditions connected with various configurations. Here we will confine ourselves only to half-planes with various boundary conditions.

If one makes $\varepsilon = \varepsilon *$, $\mu = \mu *$, and $\sigma = \sigma^*$ and supposes the boundary Σ to be a simple interface, then one gets the edge conditions for half-planes. The results obtained through the formulas given in Table 6.1 are now recapitulated in Table 6.3. It is important to observe that in such a case all the configurations are penetrable and all the results are independent of the angle θ, as expected. Observe also that only in four cases the minimal exponent ν is double root which gives rise to logarithmic singularities. We remark that the case of resistive half-plane was also considered by Braver et al. [57]. Our results differ from theirs only for E_z and H_z because according to Table 6.3 $E_z = o(1)(B_0 = 0!)$ and $H_z = O(1)(B_0 \neq 0!)$ while they give $E_z = O(1)$ and $H_z = O(r^{1/2})$.

Table 6.3.
Minimal Exponents for Half-Planes

Half-Plane	E Case	H Case
Impedance	Type: $(gg\psi\chi)$ $\nu = 1/2$ (simple) if $Z_1 Z_2 \neq 0$ $B_0 = B_0^* = 0$ if $Z_1 + Z_2 \neq 0$ $B_0 = B_0^* \neq 0$ if $Z_1 + Z_2 = 0$	Type: $(gg\psi\chi)$ $\nu = 1/2$ (simple) $B_0 = B_0^* \neq 0$ if $Z_1 + Z_2 = 0$
Resistive	Type: $(fg\psi\chi)$ $\nu = 1$(double, log·singularity) $B_0 = B_0^* = 0$ if $1/R \neq 0$	Type: $(gg\psi\chi)$ $\nu = 1/2$ (simple) if $1/R \neq 0$ $B_0 = B_0^* \neq 0$
Modified conducting	Type: $(fg\psi\chi)$ $\nu = 1$(double, log·singularity) $B_0 = B_0^* = 0$ if $1/R \neq 0$	Type: $(fg\psi\chi)$ $\nu = 1/2$ (double, log·singularity) if $1/R \neq 0$, $A \neq 0$ $B_0 = B_0^* \neq 0$ if $A \neq 0$ $B_0 = B_0^* = 0$ if $A = 0$
Magnetically conducting	Type: $(gg\psi\chi)$ $\nu = 1/2$(simple) if $1/G \neq 0$ $B_0 = B_0^* \neq 0$	Type: $(fg\psi\chi)$ $\nu = 1$ (double, log·singularity) $B_0 = B_0^* = 0$ if $1/G \neq 0$ $B_0 = B_0^* \neq 0$ if $1/G = 0$

6.6 EDGE CONDITIONS FOR THE INDUCED SURFACE CURRENTS

To conclude this part devoted to the edge singularity, we want to remark that in some numerical as well as analytical investigations one needs to know the edge behavior of the equivalent electric surface currents induced on the surface of the wedge. As is well known, the densities of these currents are defined as follows:

$$\text{On } S \qquad \mathbf{J}_S = \mathbf{n} \times [\mathbf{H}^* - \mathbf{H}], \tag{6.46a}$$

$$\text{On } \Sigma \qquad \mathbf{J}_\Sigma = \mathbf{n} \times [\mathbf{H} - \mathbf{H}^*]. \tag{6.46b}$$

The behavior of these current densities can be revealed rather easily by considering the field expressions obtained in Sections 6.3.1.B, 6.3.1.G, and 6.3.2 above. For example, from (6.16a,c) and (6.17a,c), which are valid when the wedge is penetrable but there is no confluence, one gets for the H case

$$\mathbf{J}_S = (H^*_z - H_z)\mathbf{e}_r = (k^{*2}B^*_0 - k^2 B_0)\mathbf{e}_r + O(r^\nu), \tag{6.47a}$$

$$\mathbf{J}_\Sigma = (H_z - H^*_z)\mathbf{e}_r = (k^2 B_0 - k^{*2}B^*_0)\mathbf{e}_r + O(r^\nu). \tag{6.47b}$$

From (6.47a,b) we conclude that the currents coming from S and Σ to the edge are penpendicular to the latter and, in general, individually different from zero there. However, because of the restriction (6.13a), their sum at the edge is always naught, namely,

$$(\mathbf{J}_S \cdot \mathbf{e}_r)_S + (\mathbf{J}_\Sigma \cdot \mathbf{e}_r)_\Sigma = O(r^\nu) \to 0 \qquad \text{when } r \to 0. \tag{6.48}$$

This shows that there will be no charge accumulation on the edge.

Consider now the E case, which yields

$$\mathbf{J}_S = (H_r - H^*_r)\mathbf{e}_z = O(r^{\nu-1}), \tag{6.49a}$$

$$\mathbf{J}_\Sigma = (H^*_r - H_r)\mathbf{e}_z = O(r^{\nu-1}). \tag{6.49b}$$

In contrast to the H case, now the electric current densities on S and Σ may individually become infinitely large when $r \to 0$. But, these densities are parallel to the edge and always integrable (cf. (6.14a)).

Note that all the relations (6.47a)–(6.49b) are valid when the wedge is *penetrable* and there is *no confluence* for the minimal ν. In the case of confluence, the expressions may involve some terms with $r^\nu \ln r$ or $r^{\nu-1} \ln r$, which will not change the integrability or finiteness (or infiniteness) properties of the density functions. Finally, if one replaces all the quantities having a super-index (*) by zero, then one gets the results valid for impenetrable wedges.

CHAPTER 7

Tip Singularities at the Apex of a Material Cone

7.1 INTRODUCTION

The earlier investigations connected with this type of singularity concern mainly the *algebraic* singularities that may occur at the apex of simple conical *interfaces* separating two media (see, for example, refs. 65–68). But various boundary conditions, which are considered very frequently in actual applications and studies, enforce us to consider also the *material* cones (i.e., conical sheets) and, in addition to the algebraic singularities, the logarithmic singularities (see Idemen [69]). The aim of the present section is to give some results pertinent to rather general case. To this end, one considers, without loss of generality, the tip of a rotational material cone separating the space into two regions filled with different simple materials and applies the method of *separable solutions* combined with the concept of *confluence*, which will produce also logarithmic singularities, if any.

The geometrical as well as the physical structures of the problem are schematically shown in Fig. 7.1. The space is divided into two domains, say B and B^*, by the cone S defined by $\theta = \theta_0$, and each domain is filled with certain simple material. Here (r, θ, ϕ) denote the usual spherical coordinates. The constitutive parameters of the regions B and B* are $(\varepsilon, \mu, \sigma)$ and $(\varepsilon^*, \mu^*, \sigma^*)$, respectively. In order to establish a theory as large as possible, which covers various cases corresponding to different boundary conditions on the cone, we will suppose that the cone itself has a material existence (*material cone*).

Discontinuties in the Electromagnetic Field, First Edition. By M. Mithat Idemen.

Published 2011 by John Wiley & Sons, Inc.

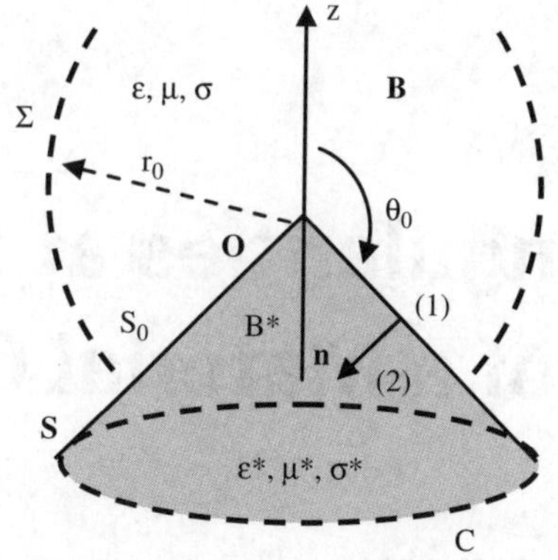

Figure 7.1. Geometry around the tip.

As such, it models a very thin conical layer about the cone $\theta = \theta_0$, that is, $\theta_0 - (\Delta\theta)/2 \leq \theta \leq \theta_0 + (\Delta\theta)/2$ (see Section 4.3.3.B).

By considering the rotational symmetry existing in the geometrical structure near the tip, we can assume without loss of generality that all the quantities connected with the field are independent of the azymuth angle ϕ. Furthermore, we can also assume that all the sources are located beyond S, inside the regions B and B^*. Thus the field inside the region B can be expressed through two scalar functions, say $U(r,\theta)$ and $V(r,\theta)$, as follows (see Jones [3, p. 486] or Harrington [63, p. 269]):

$$E_r = \frac{\partial^2}{\partial r^2}U + k'^2 U, \qquad H_r = \frac{\partial^2}{\partial r^2}V + k'^2 V \tag{7.1a}$$

$$E_\theta = \frac{1}{r}\frac{\partial^2}{\partial r \partial\theta}U, \qquad H_\theta = \frac{1}{r}\frac{\partial^2}{\partial r \partial\theta}V \tag{7.1b}$$

$$E_\phi = -i\frac{\omega\mu}{r}\frac{\partial}{\partial\theta}V, \qquad H_\phi = i\frac{\omega\varepsilon'}{r}\frac{\partial}{\partial\theta}U. \tag{7.1c}$$

Here one puts, as usual,

$$\varepsilon' = \varepsilon + i\sigma/\omega, \qquad k'^2 = \omega^2\mu\varepsilon'. \tag{7.2}$$

If in (71a)–(7.2) one replaces $\varepsilon, \varepsilon', \mu, k', U$, and V by ε^*, ε'^*, μ^*, k'^*, U^*, and V^*, then one gets the corresponding quantities in B^*.

Near the tip point (i.e. for $r \in (0, r_0)$ with a fixed $r_0 > 0$) the functions $U(r,\theta)$ and $V(r,\theta)$ satisfy the Helmholtz equations

$$\Delta(U/r) + k'^2(U/r) = 0, \quad \Delta(V/r) + k'^2(V/r) = 0, \quad \theta \neq \theta_0 \tag{7.3}$$

under the boundary conditions. Here r_0 refers to a certain constant such that the sources are located outside the sphere with radius r_0. It is worthwhile to point out that (U/r) and (V/r) appearing in (7.3) are the well-known Debye potentials while U and V themselves are the Bromwich–Borgnis functions associated with the electromagnetic field in question. In what follows we develop some general properties of U, which also concern V.

The potential function (U/r) can, in general, be expressed *for all values* of $r \in (0, \infty)$ as a linear combination of the separable solutions to the Helmholtz equation (7.3), which are of the form $U = rR_\nu(r)\Theta_\nu(\theta)$, where the separation parameters ν as well as the explicit expressions of $R_\nu(r)$ and $\Theta_\nu(\theta)$ are determined through the physical restrictions such as the boundary conditions, radiation conditions and sources. Thus, depending on whether the admissible values of ν (i.e., the *spectrum* of the problem) are discrete or continuous, one writes

$$U(r,\theta) = r \sum_{\nu \in \Im} R_\nu(r)\Theta_\nu(\theta) \tag{7.4a}$$

or

$$U(r,\theta) = r \int_\Gamma R_\nu(r)\Theta_\nu(\theta)\, d\nu. \tag{7.4b}$$

It is worthwhile to note that (7.4b) is a Kontorovich–Lebedev-type transformation when $R_\nu(r)$ is considered as the kernel [70, pp. 353–368; see also reference 71], whereas it is a Mehler–Fock-type transformation when $\Theta_\nu(\theta)$ is considered as the kernel [70, pp. 369–419]. That means that any function $U(r,\theta)$ satisfying certain integrability and continuity conditions can always be represented through an expression similar to (7.4a) or (7.4b) with appropriately chosen $\Im, \Gamma, R_\nu(r)$, and $\Theta_\nu(\theta)$. The solution of a diffraction problem under given boundary conditions consists of the determination of the set $\Im$ or the contour Γ appearing in (7.4a) or (7.4b) along with certain constants existing in the expressions of the separated functions $R_\nu(r)$ and $\Theta_\nu(\theta)$. But our aim here is quite different; we want to reveal the type and order of the possible singularity of the highest order in $U(r,\theta)$. This singularity can be determined, of course, through the function $R_\nu(r)$ at the point ν which belongs to the set $\Im$ (or the contour Γ) with the smallest real part. So we will attempt to determine this *minimal* ν by taking into account the following physical restrictions, as already mentioned in Chapter 6:

i. The energy stored in a close vicinity of the tip is finite.
ii. On the surface S the given boundary conditions are satisfied.

In what follows we will consider the issues of these restrictions. But before going into further detail, we want to recapitulate here some properties of the functions $R_\nu(r)$ and $\Theta_\nu(\theta)$ mentioned above, which will be of use in the following developments.

If we replace $U(r,\theta)$ in (7.3) by $rR(r)\Theta(\theta)$, then we get

$$r^2R'' + 2rR' + [k'^2r^2 - (\nu^2 - 1/4)]R = 0 \qquad \text{(Bessel's equation)} \tag{7.5}$$

and

$$\sin\theta(\sin\theta\Theta')' + (\nu^2 - 1/4)\sin^2\theta\Theta = 0 \qquad \text{(Legendre's equation).} \tag{7.6}$$

In (7.5) and (7.6), ν shows the same separation parameter that may have arbitrary real or complex values. Each solution of (7.5) can be written in terms of appropriate combinations of the following particular solutions (cf. (6.8a)):

$$R(r) = \begin{cases} J_\nu(k'r)/\sqrt{k'r} & \text{when } \nu \notin \{0, \pm 1, \pm 2, \ldots\} \\ J_n(k'r)/\sqrt{k'r} & \\ \quad + A_n Y_n(k'r)/\sqrt{k'r} & \text{when } \nu = n \in \{0, 1, 2, \ldots\}, \end{cases} \tag{7.7}$$

where A_n stands for an arbitrary constant while J_ν and Y_n refer to the Bessel functions of the first and second kind, respectively. Their asymptotic behaviors for $r \to 0$ are given in (6.10a,b), namely,

$$J_\nu(k'r) = c_{\nu 1}(k'r)^\nu + c_{\nu 2}(k'r)^{\nu+2} + \ldots, \quad \nu \neq -1, -2, \ldots, \tag{7.8}$$

$$Y_n(k'r) = d_{\nu 1}(k'r)^{-n} + d_{\nu 2}(k'r)^{-n+2} + \ldots + [\gamma_{\nu 1}(k'r)^n + \gamma_{\nu 2}(k'r)^{n+2} + \ldots]\ln r. \tag{7.9}$$

In (7.8) and (7.9) $c_{\nu j}, d_{\nu j}$, and $\gamma_{\nu j}\,(j = 1, 2, \ldots)$ refer to certain constants which depend also on ν and n.

As to equation (7.6), with the substitution $\cos\theta = \eta$ it is reduced to the following equation of the hypergeometric type:

$$(1-\eta^2)\frac{d^2}{d\eta^2}\Theta - 2\eta\frac{d}{d\eta}\Theta + \left(\nu^2 - \frac{1}{4}\right)\Theta = 0. \qquad (7.10)$$

This latter obviously has three regular-singular points at $\eta = 1, \eta = -1$, and $\eta = \infty$. The singularity at $\eta = 1$ (which corresponds to the semi-axis $\theta = 0$) has important issues in region B while that at $\eta = -1$ (which corresponds to the semi-axis $\theta = \pi$) plays the same role in region B^*. Therefore one has to determine the particular solutions of (7.10) such that they are bounded near the point $\eta = 1$ or $\eta = -1$. It is well known that the solution that is bounded near $\eta = 1$ consists of [72, p. 1014]

$$\Theta(\theta) = B_\nu P_{\nu-1/2}(\eta) = B_\nu F\left(\frac{1}{2} - \nu, \frac{1}{2} + \nu; 1; \frac{1-\eta}{2}\right), \qquad (7.11)$$

where B_ν stands for an arbitrary constant while $F(a, b; c; \xi)$ refers to the classical hypergeometric function defined by the series

$$F(a, b; c; \xi) = 1 + \frac{ab}{c}\xi + \frac{a(a+1)b(b+1)}{2c(c+1)}\xi^2 + \dots. \qquad (7.12)$$

This series converges in the domain $|\xi| < 1$ and, when $\xi \to 1$(i.e., when $\eta \to -1$), diverges like a logarithm. In the case of $a + b = c$, which is the case for (11), one has [73, p. 210]

$$F(a, b; c; \xi) \sim -\frac{\Gamma(a+b)}{\Gamma(a)\Gamma(b)}\ln(1-\xi). \qquad (7.13)$$

From this, one concludes that in region B the function $\Theta(\theta)$ consists of

$$\Theta(\theta) = B_\nu P_{\nu-1/2}(\cos\theta), \quad \theta \in [0, \theta_0). \qquad (7.14)$$

The expression of $\Theta(\theta)$ in region B^* is obtained rather easily by considering the fact that equation (7.10) remains unchanged under the substitution $\eta \to -\eta$. Indeed, the function $P_{\nu-1/2}(-\eta)$ is both a solution to (7.10) and bounded near $\eta = -1$. Hence, in region B^* one has

$$\Theta(\theta) = B^*{}_\nu P_{\nu-1/2}(-\cos\theta), \qquad \theta \in (\theta_0, \pi], \qquad (7.15)$$

where $B^*{}_\nu$ stands for an arbitrary constant.

In what follows we will also need the following relations that are straightforward results of (7.8), (7.9), (7.11), and (7.13): When $\theta \to \pi$, one has

$$P_{\nu-1/2}(\cos\theta) \sim -\frac{2}{\pi}\cos\nu\pi \log\left(\cos\frac{\theta}{2}\right) \tag{7.16a}$$

$$P'_{\nu-1/2}(\cos\theta) \sim \frac{d}{d(\cos\theta)}P_{\nu-1/2}(\cos\theta) \sim -\frac{1}{\pi}\frac{\cos\nu\pi}{1+\cos\theta} \tag{7.16b}$$

while when $r \to 0$ one has

$$R_\nu(r) = \begin{cases} c_{\nu 1}(k'r)^{\nu-1/2}, & \nu \neq n = \text{ integer} \\ c_{n1}(k'r)^{n-1/2} & \\ \quad + A_n d_{n1}(k'r)^{-(n+1/2)}, & \nu = n = \text{ integer} \geq 1 \\ c_{01}(k'r)^{-1/2} + \gamma_{01}A_0(\log r)/\sqrt{r}, & \nu = 0 \end{cases} \tag{7.17a}$$

and

$$\frac{d^2}{dr^2}(rR_\nu) \sim k'c_{\nu 1}(\nu^2 - 1/4)(k'r)^{\nu-3/2} + \ldots, \quad \nu \neq \pm 1/2 \tag{7.17b}$$

$$\frac{1}{r}\frac{\partial}{\partial r}(rR_\nu) \sim k'c_{\nu 1}(\nu + 1/2)(k'r)^{\nu-3/2} + \ldots, \quad \nu \neq -1/2 \tag{7.17c}$$

$$\frac{\partial^2}{\partial \mathrm{r}^2}(rR_\nu) + k'^2(rR_\nu) \equiv \frac{1}{r}(\nu^2 - 1/4)R_\nu. \tag{7.18}$$

From (7.14), (7.15), and (7.18) we conclude first of all that the separation parameter ν can never be equal to (1/2):

$$\nu \neq 1/2. \tag{7.19}$$

Indeed, for $\nu = ½$ one has

$$P_{\nu-1/2}(\pm\cos\theta) \equiv P_0(\pm\cos\theta) \equiv 1$$

and

$$\frac{\partial^2}{\partial r^2}(rR_\nu) + k'^2(rR_\nu) = 0,$$

which yield

$$\frac{\partial}{\partial\theta}U = 0, \qquad \frac{\partial^2}{\partial r^2}U + k'^2 U = 0,$$

$$\frac{\partial}{\partial\theta}V = 0, \qquad \frac{\partial^2}{\partial r^2}V + k'^2 V = 0$$

or $\mathbf{E} = 0$ and $\mathbf{H} = 0$, with (7.1a)–(7.1c) being taken into account. Therefore in our subsequent analysis we will disregard the case of $\nu = 1/2$.

It is obvious from (7.1a)–(7.1c) that in the most general case the field consists of a superposition of two simpler fields. For the first of them we have $H_r \equiv 0$ (H-type field) while for the second one we have $E_r \equiv 0$ (E-type field). For the sake of simplicity of the mathematical manipulations we can consider these particular cases separately because the results pertinent to a more general case can always be obtained by an appropriate combination of the results to be derived for them. So in what follows we will consider these cases in turn.

7.2 ALGEBRAIC SINGULARITIES OF AN *H*-TYPE FIELD

When $H_r \equiv 0$, one has $V \equiv 0$, which yields

$$E_r = \frac{\partial^2}{\partial r^2}U + k'^2 U, \qquad E_\theta = \frac{1}{r}\frac{\partial^2}{\partial r\,\partial\theta}U, \qquad H_\phi = i\frac{\omega\varepsilon'}{r}\frac{\partial}{\partial\theta}U. \tag{7.20}$$

Consider now the restrictions i and ii mentioned in Section 7.1 and try to determine the value of the minimal ν.

7.2.1 Contribution of the Energy Restriction

Since the densities of the stored electric and magnetic energies are proportional to $\mathbf{E}^2$ and $\mathbf{H}^2$, respectively, these quantities must be integrable in any region including the tip point. Therefore one has

$$\iiint_{R^3} |\mathbf{E}|^2 r^2 \sin\theta dr\, d\theta\, d\phi < \infty,$$

$$\iiint_{R^3} |\mathbf{H}|^2 r^2 \sin\theta \, dr \, d\theta \, d\phi < \infty,$$

or

$$rE_r, rE_\theta, rH_\phi \in \mathrm{L}^2{}_{\mathrm{loc}} \equiv L^2[0, r_0], \tag{7.21}$$

with the expression of the volume element in the spherical coordinate system $dV = r^2 \sin\theta dr d\theta d\phi$ being taken into account. In (7.21) r_0 stands for an arbitrary small number. By considering (7.20) and (7.17a)–(7.17c), one gets from (7.21) that

$$\Re\nu > 0, \qquad A_n \equiv 0, \qquad n = 1, 2, \ldots, \tag{7.22}$$

which, in accordance with (7.7a,b) and (7.8), yield

$$R_\nu(r) = \frac{J_\nu(k'r)}{\sqrt{k'r}} \sim c_{\nu 1}(k'r)^{\nu - 1/2} + c_{\nu 2}(k'r)^{\nu + 3/2} + \ldots. \tag{7.23}$$

If in the expressions (7.20) of the field one puts $U = rR_\nu(r)\Theta_\nu(\theta)$ and considers the relations (7.14) and (7.23), then one obtains the dominant terms of the field in the region D_1 (the following expressions are normalized by dividing with $c_{\nu 1}(k')^{\nu - 1/2}$)):

$$E_r \sim B_\nu(\nu^2 - 1/4) r^{\nu - 3/2} P_{\nu - 1/2}(\cos\theta), \tag{7.24a}$$

$$E_\theta \sim -B_\nu(\nu + 1/2) r^{\nu - 3/2} P'_{\nu - 1/2}(\cos\theta) \sin\theta, \tag{7.24b}$$

$$H_\phi \sim -B_\nu(i\omega\varepsilon'_1) r^{\nu - 1/2} P'_{\nu - 1/2}(\cos\theta) \sin\theta. \tag{7.24c}$$

Similarly, by considering (7.15) one writes the expressions valid in the region B^*:

$$E^*{}_r \sim B^*{}_\nu(\nu^2 - 1/4) r^{\nu - 3/2} P_{\nu - 1/2}(-\cos\theta), \tag{7.25a}$$

$$E^*{}_\theta \sim B^*{}_\nu(\nu + 1/2) r^{\nu - 3/2} P'_{\nu - 1/2}(-\cos\theta) \sin\theta, \tag{7.25b}$$

$$H^*{}_\phi \sim B^*{}_\nu(i\omega\varepsilon'_2) r^{\nu - 1/2} P'{}_{\nu - 1/2}(-\cos\theta) \sin\theta. \tag{7.25c}$$

7.2.2 Contribution of the Boundary Conditions

The form of the boundary conditions satisfied by the electromagnetic field on a surface S on which it suffers from discontinuity depends not only upon the constitutive parameters of the materials filling the regions separated by S but also on the topological structure and the constitutive parameters of S. As was shown in Section 4.3.3.B, the conditions in question are, in their most general form, as follows:

$$\begin{aligned}(a_1+\alpha_1 r)\mathbf{e}_\theta \times \mathbf{H}_t^{(1)} &+ (a_2+\alpha_2 r)\mathbf{e}_\theta \times \mathbf{H}_t^{(2)} \\ &+ (a_3+\alpha_3 r)\mathbf{E}_t^{(1)} + (a_4+\alpha_4 r)\mathbf{E}_t^{(2)} \\ &+ a_5\mathbf{e}_\theta \times \operatorname{grad}(rH_\theta^{(1)}) + a_6\mathbf{e}_\theta \times \operatorname{grad}(rH_\theta^{(2)}) \\ &+ a_7 \operatorname{grad}(rE_\theta^{(1)}) + a_8 \operatorname{grad}(rE_\theta^{(2)}) = 0,\end{aligned} \tag{7.26a}$$

$$\begin{aligned}(b_1+\beta_1 r)\mathbf{e}_\theta \times \mathbf{H}_t^{(1)} &+ (b_2+\beta_2 r)\mathbf{e}_\theta \times \mathbf{H}_t^{(2)} \\ &+ (b_3+\beta_3 r)\mathbf{E}_t^{(1)} + (b_4+\beta_4 r)\mathbf{E}_t^{(2)} \\ &+ b_5\mathbf{e}_\theta \times \operatorname{grad}(rH_\theta^{(1)}) + b_6\mathbf{e}_\theta \times \operatorname{grad}(rH_\theta^{(2)}) \\ &+ b_7 \operatorname{grad}(rE_\theta^{(1)}) + b_8 \operatorname{grad}(rE_\theta^{(2)}) = 0.\end{aligned} \tag{7.26b}$$

Here a_j, α_j, b_j, and β_j refer to certain complex quantities that can be determined by considering the thin conical layer modeled by S as well as the materials filling the regions B and B^*. They satisfy the identities (4.72b,c). The super-indices (1) and (2) appearing in $E^{1,2}$ and $H^{1,2}$ indicate, as usual, that these quantities are the limiting values that are observed when one approaches the surface S from the regions B and B^*, respectively. As to the sub-indices t and θ, they refer to the projections onto the tangent plane and normal to S at the point in question, respectively.

In the present H case we have

$$H^{1,2}{}_\theta = 0, \qquad \mathbf{H}^{1,2}{}_t = H^{1,2}{}_\phi \mathbf{e}_\phi, \qquad \mathbf{E}^{1,2}{}_t = E^{1,2}{}_r \mathbf{e}_r$$

while the dominant terms of the asymptotic expressions of $H^{1,2}{}_\phi, E^{1,2}{}_r$, and $E^{1,2}{}_\theta$ for $r \to 0$ are given by (7.24a–c). If one puts these latter equations into (7.26a,b), then one gets two equations involving the terms of orders $r^{\nu-3/2}, r^{\nu-1/2}$ and $r^{\nu+1/2}$. By equating the coefficients of the least-order terms in each of these equations to zero, one obtains two equations that will be used to determine the minimal value of ν. Since $\nu \neq 1/2$ (see (7.19)), (7.26a) yields the following three possibilities:

Case C_1: If $|a_3| + |a_4| + |a_7| + |a_8| \neq 0$, then one has (for the terms of order $r^{\nu-3/2}$)

$$[a_3 P_{\nu-1/2}(\cos\theta_0) - a_7 P'_{\nu-1/2}(\cos\theta_0)\sin\theta_0]B_\nu + [a_4 P_{\nu-1/2}(-\cos\theta_0) + a_8 P'_{\nu-1/2}(-\cos\theta_0)\sin\theta_0]B^*_\nu = 0. \quad (7.27a)$$

Case C_2: If $|a_3| + |a_4| + |a_7| + |a_8| = 0$ while $|a_1| + |a_2| + |\alpha_3| + |\alpha_4| \neq 0$, then one has (for the terms of order $r^{\nu-1/2}$)

$$[-i\omega\varepsilon'_1 a_1 P'_{\nu-1/2}(\cos\theta_0)\sin\theta_0 + \alpha_3(\nu^2-1/4)P_{\nu-1/2}(\cos\theta_0)]B_\nu + [i\omega\varepsilon'_2 a_2 P'_{\nu-1/2}(-\cos\theta_0)\sin\theta_0 + \alpha_4(\nu^2-1/4)P_{\nu-1/2}(-\cos\theta_0)]B^*_\nu = 0. \quad (7.27b)$$

Case C_3: If $|a_3| + |a_4| + |a_7| + |a_8| = 0$ and $|a_1| + |a_2| + |\alpha_3| + |\alpha_4| = 0$ while $|\alpha_1| + |\alpha_2| \neq 0$, then one has (for the terms of order $r^{\nu+1/2}$)

$$[\varepsilon'_1 \alpha_1 P'_{\nu-1/2}(\cos\theta_0)]B_\nu - [\varepsilon'_2 \alpha_2 P'_{\nu-1/2}(-\cos\theta_0)]B^*_\nu = 0. \quad (7.27c)$$

The cases and equations pertinent to the second boundary condition (7.26b) are obtained from the above-mentioned ones by making the substitutions $a_j \to b_j$ and $\alpha_j \to \beta_j$. In what follows, these cases will be denoted by C^*_1, C^*_2, and C^*_3. Thus all the possible situations become consist of the following nine *canonical* cases:

$$\begin{matrix} (C_1, C^*_1), & (C_1, C^*_2), & (C_1, C^*_3), \\ (C_2, C^*_1), & (C_2, C^*_2), & (C_2, C^*_3), \\ (C_3, C^*_1), & (C_3, C^*_2), & (C_3, C^*_3). \end{matrix}$$

It is obvious that only six of these cases may result in mathematically different problems. They are

$$\begin{matrix} (C_1, C^*_1), & (C_1, C^*_2), & (C_1, C^*_3), \\ (C_2, C^*_2), & (C_2, C^*_3), & (C_3, C^*_3). \end{matrix}$$

Observe that the resulting equations are nothing but linear and homogeneous algebraic equations with respect to the unknown coefficients B_ν and B^*_ν. Thus the compatibility condition of the equations pertinent to the case (C_i, C^*_j) results in a transcendental equation in terms of ν, say $\Delta_{ij}(\nu) = 0$, which permit us to determine the admissible values of v. The explicit expressions of the functions (determinants) $\Delta_{ij}(v)$ are as follows:

$$\Delta_{ij}(v) = \begin{vmatrix} A_{ij} & B_{ij} \\ C_{ij} & D_{ij} \end{vmatrix}, \qquad i,j = 1,2,3$$

where

$$\begin{aligned}
A_{11} &= a_3 P_{v-1/2}(\cos\theta_0) - a_7 P'_{v-1/2}(\cos\theta_0)\sin\theta_0 \\
B_{11} &= a_4 P_{v-1/2}(-\cos\theta_0) + a_8 P'_{v-1/2}(-\cos\theta_0)\sin\theta_0 \\
C_{11} &= b_3 P_{v-1/2}(\cos\theta_0) - b_7 P'_{v-1/2}(\cos\theta_0)\sin\theta_0 \\
D_{11} &= b_4 P_{v-1/2}(-\cos\theta_0) + b_8 P'_{v-1/2}(-\cos\theta_0)\sin\theta_0
\end{aligned}$$

$$\begin{aligned}
A_{12} &= a_3 P_{v-1/2}(\cos\theta_0) - a_7 P'_{v-1/2}(\cos\theta_0)\sin\theta_0 \\
B_{12} &= a_4 P_{v-1/2}(-\cos\theta_0) + a_8 P'_{v-1/2}(-\cos\theta_0)\sin\theta_0 \\
C_{12} &= \beta_3(v^2-1/4)P_{v-1/2}(\cos\theta_0) - b_1(i\omega\varepsilon'_1)P'_{v-1/2}(\cos\theta_0)\sin\theta_0 \\
D_{12} &= \beta_4(v^2-1/4)P_{v-1/2}(-\cos\theta_0) \\
&\qquad + b_2(i\omega\varepsilon'_2)P'_{v-1/2}(-\cos\theta_0)\sin\theta_0
\end{aligned}$$

$$\begin{aligned}
A_{13} &= a_3 P_{v-1/2}(\cos\theta_0) - a_7 P'_{v-1/2}(\cos\theta_0)\sin\theta_0 \\
B_{13} &= a_4 P_{v-1/2}(-\cos\theta_0) + a_8 P'_{v-1/2}(-\cos\theta_0)\sin\theta_0 \\
C_{13} &= \beta_1\varepsilon'_1 P'_{v-1/2}(\cos\theta_0) \\
D_{13} &= -\beta_2\varepsilon'_2 P'_{v-1/2}(-\cos\theta_0)
\end{aligned}$$

$$\begin{aligned}
A_{22} &= \alpha_3(v^2-1/4)P_{v-1/2}(\cos\theta_0) - a_1(i\omega\varepsilon'_1)P'_{v-1/2}(\cos\theta_0)\sin\theta_0 \\
B_{22} &= \alpha_4(v^2-1/4)P_{v-1/2}(-\cos\theta_0) + a_2(i\omega\varepsilon'_2)P'_{v-1/2}(-\cos\theta_0)\sin\theta_0 \\
C_{22} &= \beta_3(v^2-1/4)P_{v-1/2}(\cos\theta_0) - b_1(i\omega\varepsilon'_1)P'_{v-1/2}(\cos\theta_0)\sin\theta_0 \\
D_{22} &= \beta_4(v^2-1)P_{v-1/2}(-\cos\theta_0) + b_2(i\omega\varepsilon'_2)P'_{v-1/2}(-\cos\theta_0)\sin\theta_0
\end{aligned}$$

$$\begin{aligned}
A_{23} &= \alpha_3(v^2-1/4)P_{v-1/2}(\cos\theta_0) - a_1(i\omega\varepsilon'_1)P'_{v-1/2}(\cos\theta_0)\sin\theta_0 \\
B_{23} &= \alpha_4(v^2-1/4)P_{v-1/2}(-\cos\theta_0) + a_2(i\omega\varepsilon'_2)P'_{v-1/2}(-\cos\theta_0)\sin\theta_0 \\
C_{23} &= \beta_1\varepsilon'_1 P'_{v-1/2}(\cos\theta_0) \\
D_{23} &= -\beta_2\varepsilon'_2 P'_{v-1/2}(-\cos\theta_0)
\end{aligned}$$

$$\begin{aligned}
A_{33} &= \alpha_1 P'_{v-1/2}(\cos\theta_0)P'_{v-1/2}(-\cos\theta_0) \\
B_{33} &= \alpha_2 P'_{v-1/2}(\cos\theta_0)P'_{v-1/2}(-\cos\theta_0) \\
C_{33} &= \beta_1 P'_{v-1/2}(\cos\theta_0)P'_{v-1/2}(-\cos\theta_0) \\
D_{33} &= \beta_2 P'_{v-1/2}(\cos\theta_0)P'_{v-1/2}(-\cos\theta_0).
\end{aligned}$$

Note that the compatibility conditions corresponding to other cases can be obtained from what are given above by simple straightforward substitutions, namely,

$$\Delta_{21}(\nu) = \Delta_{12}(\nu, a_j \leftrightarrow b_j, \alpha_j \leftrightarrow \beta_j),$$
$$\Delta_{31}(\nu) = \Delta_{13}(\nu, a_j \leftrightarrow b_j, \alpha_j \leftrightarrow \beta_j),$$
$$\Delta_{32}(\nu) = \Delta_{23}(\nu, a_j \leftrightarrow b_j, \alpha_j \leftrightarrow \beta_j).$$

Example As an example consider the case when S consists of a *resistive* boundary (see (4.73g)) on which one has

$$\mathbf{E}_{1t} - \mathbf{E}_{2t} = 0, \quad \mathbf{e}_\theta \times (\mathbf{H}_{1t} - \mathbf{H}_{2t}) + \frac{1}{2R} r(\mathbf{E}_{1t} + \mathbf{E}_{2t}) = 0. \quad (7.28a)$$

Here R stands for a given positive constant while $\mathbf{E}_{1,2t} = E_{1,2r}\mathbf{e}_r$ and $\mathbf{H}_{1,2t} = H_{1,2\phi}\mathbf{e}_\phi$. The boundary condition table corresponding to (7.28a), which satisfies also (4.72b,c), (4.77) and (4.78), is as follows:

$$C = \begin{bmatrix} 0 & 0 & 1 & -1 & 0 & 0 & 0 & 0 \\ 0 & 0 & 0 & 0 & - & - & - & - \\ 1 & -1 & 0 & 0 & 0 & 0 & 0 & 0 \\ 0 & 0 & 1/(2R) & 1/(2R) & - & - & - & - \end{bmatrix}.$$

Since $|a_3| + |a_4| + |a_7| + |a_8| \neq 0$ while $|b_3| + |b_4| + |b_7| + |b_8| = 0$ but $|b_1| + |b_2| + |\beta_3| + |\beta_4| \neq 0$, the pertinent canonical case is $(C_1, C_2{}^*)$ which requires $\Delta_{12}(\nu) = 0$; namely:

$$\begin{aligned}(\nu^2 - 1/4)&P_{\nu-1/2}(\cos\theta_0)P_{\nu-1/2}(-\cos\theta_0)\\ &- i\omega R[\varepsilon_1' P_{\nu-1/2}(-\cos\theta_0)P'_{\nu-1/2}(\cos\theta_0)\\ &+ \varepsilon_2' P_{\nu-1/2}(\cos\theta_0)P'_{\nu-1/2}(-\cos\theta_0)]\sin\theta_0 = 0 \quad (7.28b)\end{aligned}$$

The solution to this equation with minimum positive real part ($\nu \neq 1/2$) will give us the minimal ν in question (see Section 7.6 below). It is obvious that in this particular case of polarization and boundary condition, the minimal value of ν depends not only on the apex angle θ_0 but also on the constitutive parameters of the space and the value of the resistivity of the cone.

7.3 ALGEBRAIC SINGULARITIES OF AN E-TYPE FIELD

When $U \equiv 0$, the equations (7.1a–c) define a field with $E_r = 0$.The expressions pertinent to this field are obtained from what are connected with the aforementioned H case by making the following substitutions:

$$U \to V, \qquad \varepsilon' \to -\mu, \qquad E_{r,\theta} \to H_{r,\theta}, \qquad H_\phi \to E_\phi. \tag{7.29}$$

One has, for example,

$$H_r \sim B_\nu(\nu^2 - 1/4)r^{\nu-3/2}P_{\nu-1/2}(\cos\theta), \tag{7.30a}$$

$$H_\theta \sim -B_\nu(\nu + 1/2)r^{\nu-3/2}P'_{\nu-1/2}(\cos\theta)\sin\theta, \tag{7.30b}$$

$$E_\phi \sim B_\nu(i\omega\mu)r^{\nu-1/2}P'_{\nu-1/2}(\cos\theta)\sin\theta, \tag{7.30c}$$

$$H_r^* \sim B_\nu^*(\nu^2 - 1/4)r^{\nu-3/2}P_{\nu-1/2}(-\cos\theta), \tag{7.30d}$$

$$H_\theta^* \sim B_\nu^*(\nu + 1/2)r^{\nu-3/2}P'_{\nu-1/2}(-\cos\theta)\sin\theta, \tag{7.30e}$$

$$E_\phi^* \sim -B_\nu^*(i\omega\mu^*)r^{\nu-1/2}P'_{\nu-1/2}(-\cos\theta)\sin\theta. \tag{7.30f}$$

On the other hand, owing to the equalities

$$E^{1,2}{}_\theta = 0, \qquad \mathbf{H}^{1,2}{}_t = H^{1,2}{}_r\mathbf{e}_r, \qquad \mathbf{E}^{1,2}{}_t = E^{1,2}{}_\phi\mathbf{e}_\phi,$$

the boundary condition (7.26a) becomes

$$(a_1 + \alpha_1 r)H^1{}_r + (a_2 + \alpha_2 r)H^2{}_r - (a_3 + \alpha_3 r)E^1{}_\phi - (a_4 + \alpha_4 r)E^2{}_\phi$$
$$+ a_5\frac{\partial}{\partial r}(rH^1{}_\theta) + a_6\frac{\partial}{\partial r}(rH^2{}_\theta) = 0. \tag{7.31}$$

A similar equation is also written with a_j and α_j replaced by b_j and β_j, respectively. If one compares these equations with those pertinent to the H case, one observes readily that the relations connected with the E case can be obtained from what are pertinent to the H case by making simply the substitutions:

$$\begin{aligned}
&a_3 \leftrightarrow a_1, \quad a_7 \to a_5, \quad a_4 \leftrightarrow a_2, \quad a_8 \to a_6,\\
&\alpha_3 \leftrightarrow \alpha_1, \quad \alpha_4 \leftrightarrow \alpha_2, \quad \varepsilon'_1 \to \mu,\\
&b_3 \leftrightarrow b_1, \quad b_7 \to b_5, \quad b_4 \leftrightarrow b_2, \quad b_8 \to b_6,\\
&\beta_3 \leftrightarrow \beta_1, \quad \beta_4 \leftrightarrow \beta_2, \quad \varepsilon'_2 \to \mu^*.
\end{aligned}$$

Thus the three possible E cases are as follows:

Case C_1: $|a_1| + |a_2| + |a_5| + |a_6| \neq 0$.

Case C_2: $|a_1| + |a_2| + |a_5| + |a_6| = 0$ but
$|a_3| + |a_4| + |\alpha_1| + |\alpha_2| \neq 0$.

Case C_3: $|a_1| + |a_2| + |a_5| + |a_6| = 0$ and
$|a_3| + |a_4| + |\alpha_1| + |\alpha_2| = 0$ but $|\alpha_3| + |\alpha_4| \neq 0$.

The determinants Δ_{ij} pertinent to these canonical cases are as follows

$$\Delta_{ij}(\nu) = \begin{vmatrix} A_{ij} & B_{ij} \\ C_{ij} & D_{ij} \end{vmatrix}, \qquad i,j = 1,2,3$$

where

$$\begin{aligned}
A_{11} &= a_1 P_{\nu-1/2}(\cos\theta_0) - a_5 P'_{\nu-1/2}(\cos\theta_0)\sin\theta_0 \\
B_{11} &= a_2 P_{\nu-1/2}(-\cos\theta_0) + a_6 P'_{\nu-1/2}(-\cos\theta_0)\sin\theta_0 \\
C_{11} &= b_1 P_{\nu-1/2}(\cos\theta_0) - b_5 P'_{\nu-1/2}(\cos\theta_0)\sin\theta_0 \\
D_{11} &= b_2 P_{\nu-1/2}(-\cos\theta_0) + b_6 P'_{\nu-1/2}(-\cos\theta_0)\sin\theta_0
\end{aligned}$$

$$\begin{aligned}
A_{12} &= a_1 P_{\nu-1/2}(\cos\theta_0) - a_5 P'_{\nu-1/2}(\cos\theta_0)\sin\theta_0 \\
B_{12} &= a_2 P_{\nu-1/2}(-\cos\theta_0) + a_6 P'_{\nu-1/2}(-\cos\theta_0)\sin\theta_0 \\
C_{12} &= \beta_1(\nu^2-1/4)P_{\nu-1/2}(\cos\theta_0) - b_3(i\omega\mu)P'_{\nu-1/2}(\cos\theta_0)\sin\theta_0 \\
D_{12} &= \beta_2(\nu^2-1/4)P_{\nu-1/2}(-\cos\theta_0) \\
&\qquad + b_4(i\omega\mu^*)P'_{\nu-1/2}(-\cos\theta_0)\sin\theta_0
\end{aligned}$$

$$\begin{aligned}
A_{13} &= a_1 P_{\nu-1/2}(\cos\theta_0) - a_5 P'_{\nu-1/2}(\cos\theta_0)\sin\theta_0 \\
B_{13} &= a_2 P_{\nu-1/2}(-\cos\theta_0) + a_6 P'_{\nu-1/2}(-\cos\theta_0)\sin\theta_0 \\
C_{13} &= \beta_3 \mu P'_{\nu-1/2}(\cos\theta_0) \\
D_{13} &= -\beta_4 \mu^* P'_{\nu-1/2}(-\cos\theta_0)
\end{aligned}$$

$$\begin{aligned}
A_{22} &= \alpha_1(\nu^2-1/4)P_{\nu-1/2}(\cos\theta_0) - a_3(i\omega\mu)P'_{\nu-1/2}(\cos\theta_0)\sin\theta_0 \\
B_{22} &= \alpha_2(\nu^2-1/4)P_{\nu-1/2}(-\cos\theta_0) \\
&\qquad + a_4(i\omega\mu^*)P'_{\nu-1/2}(-\cos\theta_0)\sin\theta_0 \\
C_{22} &= \beta_1(\nu^2-1/4)P_{\nu-1/2}(\cos\theta_0) - b_3(i\omega\mu)P'_{\nu-1/2}(\cos\theta_0)\sin\theta_0 \\
D_{22} &= \beta_2(\nu^2-1/4)P_{\nu-1/2}(-\cos\theta_0) \\
&\qquad + b_4(i\omega\mu^*)P'_{\nu-1/2}(-\cos\theta_0)\sin\theta_0
\end{aligned}$$

$$\begin{aligned}
A_{23} &= \alpha_1(\nu^2-1/4)P_{\nu-1/2}(\cos\theta_0) - a_3(i\omega\mu)P'_{\nu-1/2}(\cos\theta_0)\sin\theta_0 \\
B_{23} &= \alpha_2(\nu^2-1/4)P_{\nu-1/2}(-\cos\theta_0) \\
&\qquad + a_4(i\omega\mu^*)P'_{\nu-1/2}(-\cos\theta_0)\sin\theta_0 \\
C_{23} &= \beta_3 \mu P'_{\nu-1/2}(\cos\theta_0) \\
D_{23} &= -\beta_4 \mu^* P'_{\nu-1/2}(-\cos\theta_0)
\end{aligned}$$

$$A_{33} = \alpha_3 P'_{\nu-1/2}(\cos\theta_0)P'_{\nu-1/2}(-\cos\theta_0)$$
$$B_{33} = \alpha_4 P'_{\nu-1/2}(\cos\theta_0)P'_{\nu-1/2}(-\cos\theta_0)$$
$$C_{33} = \beta_3 P'_{\nu-1/2}(\cos\theta_0)P'_{\nu-1/2}(-\cos\theta_0)$$
$$D_{33} = \beta_4 P'_{\nu-1/2}(\cos\theta_0)P'_{\nu-1/2}(-\cos\theta_0).$$

Note that the compatibility conditions corresponding to other cases can be obtained from what are given above by simple straightforward substitutions, namely,

$$\Delta_{21}(\nu) = \Delta_{12}(\nu, a_j \leftrightarrow b_j, \alpha_j \leftrightarrow \beta_j),$$
$$\Delta_{31}(\nu) = \Delta_{13}(\nu, a_j \leftrightarrow b_j, \alpha_j \leftrightarrow \beta_j),$$
$$\Delta_{32}(\nu) = \Delta_{23}(\nu, a_j \leftrightarrow \mathrm{b}_j, \alpha_j \leftrightarrow \beta_j).$$

Example Consider again the case of the *resistive* boundary on which one has (7.28a), where

$$\mathbf{H}_{1,2t} = \mathrm{H}_{1,2r}\ \mathbf{e}_r, \quad \mathbf{E}_{1,2t} = \mathrm{E}_{1,2\phi}\ \mathbf{e}_\phi.$$

Since one has has now

$$|a_1| + |a_2| + |a_5| + |a_6| = 0, \quad |a_3| + |a_4| + |\alpha_1| + |\alpha_2| \neq 0$$

and

$$|b_1| + |b_2| + |b_5| + |b_6| \neq 0,$$

the pertinent canonical case is $(C_2, C_1{}^*)$ which requires $\Delta_{21}(\nu) = 0$, namely:

$$\mu P'_{\nu-1/2}(\cos\theta_0)P_{\nu-1/2}(-\cos\theta_0)$$
$$+ \mu^* P'_{\nu-1/2}(-\cos\theta_0)P_{\nu-1/2}(\cos\theta_0) = 0. \qquad (7.32)$$

The value of the minimal ν is the solution to this transcendental equation with minimum positive real part (see Section 7.6 below). It is obviously depend on both the constitutive parameters of the space and the apex angle.

7.4 THE CASE OF IMPENETRABLE CONES

If the coefficients with even indices in the equation (7.26a) are all zero, then the boundary S becomes *impenetrable* in point of the region B.

In such a case the sources located *outside* B have no effect on the field observed inside B. Since in this case the equations (7.27a–c) do not involve $B^*{}_\nu$, the compatibility conditions are reduced to the equations obtained by equating to zero the coefficients of B_ν. Thus the equations from which one finds the minimal value of ν become the following:

In H Case

Case C_1: If $|a_3| + |a_7| \neq 0$, then one has (for the terms of order $r^{\nu-3/2}$)

$$a_3 P_{\nu-1/2}(\cos\theta_0) - a_7 P'_{\nu-1/2}(\cos\theta_0)\sin\theta_0 = 0. \tag{7.33a}$$

Case C_2: If $|a_3| + |a_7| = 0$ while $|a_1| + |\alpha_3| \neq 0$, then one has (for the terms of order $r^{\nu-1/2}$)

$$\begin{aligned} &- i\omega\varepsilon'_1 a_1 P'_{\nu-1/2}(\cos\theta_0)\sin\theta_0 \\ &+ \alpha_3(\nu^2 - 1/4)P_{\nu-1/2}(\cos\theta_0) = 0. \end{aligned} \tag{7.33b}$$

Case C_3: If $|a_3| + |a_7| = 0, |a_1| + |\alpha_3| = 0$ while $\alpha_1 \neq 0$, then one has (for the terms of order $r^{\nu+1/2}$)

$$P'_{\nu-1/2}(\cos\theta_0) = 0. \tag{7.33c}$$

In E Case

Case C_1: If $|a_1| + |a_5| \neq 0$, then one has (for the terms of order $r^{\nu-3/2}$)

$$a_1 P_{\nu-1/2}(\cos\theta_0) - a_5 P'_{\nu-1/2}(\cos\theta_0)\sin\theta_0 = 0. \tag{7.34a}$$

Case C_2: If $|a_1| + |a_5| = 0$ while $|a_3| + |\alpha_1| \neq 0$, then one has (for the terms of order $r^{\nu-1/2}$)

$$\begin{aligned} &i\omega\mu a_3 P'_{\nu-1/2}(\cos\theta_0)\sin\theta_0 \\ &- \alpha_1(\nu^2 - 1/4)P_{\nu-1/2}(\cos\theta_0) = 0. \end{aligned} \tag{7.34b}$$

Case C_3: If $|a_1| + |a_5| = 0, |a_3| + |\alpha_1| = 0$ while $\alpha_3 \neq 0$, then one has (for the terms of order $r^{\nu+1/2}$)

$$P'_{\nu-1/2}(\cos\theta_0) = 0. \tag{7.34c}$$

Example. As an example consider the case when S consists of a perfectly conducting boundary. In this case one has (see (4.90))

$$r(\mathbf{E}_{1t} + \mathbf{E}_{2t}) = 0, \qquad \mathbf{E}_{1t} - \mathbf{E}_{2t} = 0, \tag{7.35a}$$

The boundary condition table corresponding to (7.35a), which satisfies also (4.72b,c), (4.77) and (4.78), is as follows:

$$C = \begin{bmatrix} 0 & 0 & 0 & 0 & 0 & 0 & 0 & 0 \\ 0 & 0 & 1 & 1 & - & - & - & - \\ 0 & 0 & 1 & -1 & 0 & 0 & 0 & 0 \\ 0 & 0 & 0 & 0 & - & - & - & - \end{bmatrix}. \tag{7.35b}$$

From this table we get

$$|a_3| + |a_7| = 0 \text{ but } |a_1| + |\alpha_3| \neq 0$$

while

$$|a_1| + |a_5| = 0, \ |a_3| + |\alpha_1| = 0 \text{ but } \alpha_3 \neq 0.$$

Thus from (7.33b) and (7.34 c) one concludes that the pertinent canonical case is C_2 for the H-case and C_3 for the E-case. Therefore the minimal value of ν is determined through the following rather simple equations:

$$\text{in the } H\text{-case:} \quad P_{\nu-1/2}(\cos\theta_0) = 0, \tag{7.35c}$$

$$\text{in the } E\text{-case:} \quad P'_{\nu-1/2}(\cos\theta_0) = 0. \tag{7.35d}$$

This shows that ν does not depend on the constitutive parameters of the space; it is determined only through the apex angle θ_0 (for the determination of ν see Section 7.6 below).

7.5 CONFLUENCE AND LOGARITHMIC SINGULARITIES

All the formulas derived above were based on simple separable expressions that tacitly assumed that the minimal point ν is a *simple* root of the equation $\Delta_{ij}(\nu) = 0$. If the situation is not so, i.e. when ν is a multiple zero of the function $\Delta_{ij}(\nu)$, then certain additional terms including $\log r$ will also take place in the expressions in question. Now we want to consider this case (see Idemen [69]).

The multiplicity of v can be thought as a result of certain type of *coalescence* [69]. To clarify the idea and the method, we will consider here the simplest case where v consists of a double zero of $\Delta_{ij}(v)$. If some of the parameters—for example, ε'_1—change slightly to become $\varepsilon'_1 + \Delta\varepsilon$, then the zeros of $\Delta_{ij}(v)$, which are located in a close vicinity of the point v, also change and may become simple. We will denote them by v' and v''. When $\Delta\varepsilon \to 0$, the coalescence $v'' \to v' \to v$ generates the double zero v. Now we will briefly show some issues of this process.

Consider first the expression of E_r given by (7.24a) in the case of $\varepsilon'_1 + \Delta\varepsilon$ mentioned above. When we consider both of the zeros that will tend to v as $\Delta\varepsilon \to 0$, then we write

$$E_r \sim B_1(v')(v'^2 - 1/4)r^{v'-3/2}P_{v'-1/2}(\cos\theta) \\ + B_2(v'')(v''^2 - 1/4)r^{v''-3/2}P_{v''-1/2}(\cos\theta), \quad (7.36)$$

where $B_1(v')$ and $B_2(v'')$ refer to certain coefficients. If these latter individually tend to certain *finite* limits when $v'' \to v' \to v$, then (7.36) gives only (7.24a). But this is not the only possibility that produces a finite limit. Indeed, if $B_1(v')$ and $B_2(v'')$ tend individually to *infinity* such that

$$B_1(v') \to B_1 + C_1/(v' - v'')$$

and

$$B_2(v'') \to B_2 + C_1/(v'' - v'),$$

where B_1, B_2, and C_1 refer to certain constants, then (7.36) will also produce a finite limit, namely,

$$E_r \sim Br^{v-3/2}P_{v-1/2}(\cos\theta) + C\frac{\partial}{\partial v}\{r^{v-3/2}P_{v-1/2}(\cos\theta)\} \\ \sim Br^{v-3/2}P_{v-1/2}(\cos\theta) + Cr^{v-3/2}[\log rP_{v-1/2}(\cos\theta) \\ + \frac{\partial}{\partial v}P_{v-1/2}(\cos\theta)]. \quad (7.37)$$

Here B and C stand for certain constants. The first term in (7.37) is quite similar to that in (7.24a) but the additional one is the result of the confluence. So we can refer the expression in (7.37) as to be the *confluent* asymptotic expression of E_r. Note that the additional part in (7.37) is not in separated form as the first one. In other words, the separated expressions (7.24a)–(7.25c) cannot individually involve logarithmic singularities. Such a singularity in the field may come from certain kind of *confluence* that generates *nonseparable* terms.

By repeating the above reasoning, we reveal the confluent expressions pertinent to the other field components. For the sake of completeness, we write down here the explicit confluent expressions pertinent to (7.24a)–(7.24c):

$$\begin{aligned}
E_r \sim{} & [2B(\nu^2 - 1/4) + 2C\nu]r^{\nu-3/2}P_{\nu-1/2}(\cos\theta) \\
& + C(\nu^2 - 1/4)r^{\nu-3/2}[\log r P_{\nu-1/2}(\cos\theta) \\
& \qquad + \frac{\partial}{\partial\nu}P_{\nu-1/2}(\cos\theta)], \\
E_\theta \sim{} & -[2B(\nu + 1/2) + C]r^{\nu-3/2}P'_{\nu-1/2}(\cos\theta)\sin\theta \\
& - C(\nu + 1/2)r^{\nu-3/2}[\log r P'_{\nu-1/2}(\cos\theta) \\
& \qquad + \frac{\partial}{\partial\nu}P'_{\nu-1/2}(\cos\theta)]\sin\theta \\
H_\phi \sim{} & -2B(i\omega\varepsilon'_1)r^{\nu-1/2}P'_{\nu-1/2}(\cos\theta)\sin\theta \\
& - C(i\omega\varepsilon'_1)r^{\nu-1/2}[\log r P'_{\nu-1/2}(\cos\theta) \\
& \qquad + \frac{\partial}{\partial\nu}P'_{\nu-1/2}(\cos\theta)]\sin\theta.
\end{aligned}$$

The expressions valid for the field in B^* as well as those connected with the E case are written similarly. For a numerical example pertinent to this case see Section 7.7.2 below.

7.6 APPLICATION TO SOME WIDELY USED ACTUAL BOUNDARY CONDITIONS

In the most important actual applications the boundaries consist of simple interfaces or perfectly conducting, resistive, magnetically conducting etc boundaries. The cases of resistive and perfectly conducting boundaries were already considered as examples in Sections 7.2.2, 7.3 and 7.4 above. Now we want to briefly consider the other cases. The boundary conditions as well as the boundary condition tables pertinent to these cases are as follows:

Simple interface (see (4.87)):

$$\mathbf{E}_{1t} - \mathbf{E}_{2t} = 0, \qquad \mathbf{e}_\theta \times (\mathbf{H}_{1t} - \mathbf{H}_{2t}) = 0 \Rightarrow$$

$$C = \begin{bmatrix} 0 & 0 & 1 & -1 & 0 & 0 & 0 & 0 \\ 0 & 0 & 0 & 0 & - & - & - & - \\ 1 & -1 & 0 & 0 & 0 & 0 & 0 & 0 \\ 0 & 0 & 0 & 0 & - & - & - & - \end{bmatrix}.$$

Magnetically conducting sheet (see (4.89)):

$$\mathbf{E}_{1t} - \mathbf{E}_{2t} - r\frac{1}{2G}\mathbf{e}_\theta \times (\mathbf{H}_{1t} + \mathbf{H}_{2t}) = 0, \quad \mathbf{e}_\theta \times (\mathbf{H}_{1t} - \mathbf{H}_{2t}) = 0 \Rightarrow$$

$$C = \begin{bmatrix} 0 & 0 & 1 & -1 & 0 & 0 & 0 & 0 \\ -1/(2G) & -1/(2G) & 0 & 0 & - & - & - & - \\ 1 & -1 & 0 & 0 & 0 & 0 & 0 & 0 \\ 0 & 0 & 0 & 0 & - & - & - & - \end{bmatrix}.$$

By considering the above-mentioned tables together with the definitions of the cases C_1, C_2, and C_3 mentioned in Sections 7.2, 7.3, and 7.4, we convince ourselves that the above-mentioned boundary conditions give rise to the cases shown in Table 7.1 (for the sake of completeness we include into Table 7.1 the previously studied simple, resistive and perfectly conducting cases also).

Table 7.1.
(Cases Pertinent to Different Boundaries)

Boundary	Polarization	
	H-Case	*E*-Case
Simple interface (penetrable)	(C_1, C^*_2)	(C_2, C^*_1)
Resistive boundary (penetrable)	(C_1, C^*_2)	(C_2, C^*_1)
Magnetically conducting boundary (penetrable)	(C_1, C^*_2)	(C_2, C^*_1)
Perfectly conducting boundary (impenetrable)	(C_2)	(C_3)

If one considers the appropriate equations $\Delta_{ij}(\nu) = 0$, from the Table 7.1 one gets the following transcendental equations which permit us to determine the minimal values of ν (Note that because of (7.19) $\nu \neq 1/2$):

Simple interface:

$$\begin{aligned} H\text{-case} \Rightarrow\ & \varepsilon_1' P_{\nu-1/2}'(\cos\theta_0) P_{\nu-1/2}(-\cos\theta_0) \\ & + \varepsilon_2' P_{\nu-1/2}'(-\cos\theta_0) P_{\nu-1/2}(\cos\theta_0) = 0, \\ E\text{-case} \Rightarrow\ & \mu P_{\nu-1/2}'(\cos\theta_0) P_{\nu-1/2}(-\cos\theta_0) \\ & + \mu^* P_{\nu-1/2}'(-\cos\theta_0) P_{\nu-1/2}(\cos\theta_0) = 0. \end{aligned}$$

Resistive boundary:

$$\begin{aligned} H\text{-case} \Rightarrow\ & (\nu^2 - 1/4) P_{\nu-1/2}(\cos\theta_0) P_{\nu-1/2}(-\cos\theta_0) \\ & - i\omega R[\varepsilon_1' P_{\nu-1/2}(-\cos\theta_0) P_{\nu-1/2}'(\cos\theta_0) \\ & + \varepsilon_2' P_{\nu-1/2}(\cos\theta_0) P_{\nu-1/2}'(-\cos\theta_0)] \sin\theta_0 = 0 \\ E\text{-case} \Rightarrow\ & \mu P_{\nu-1/2}'(\cos\theta_0) P_{\nu-1/2}(-\cos\theta_0) \\ & + \mu^* P_{\nu-1/2}'(-\cos\theta_0) P_{\nu-1/2}(\cos\theta_0) = 0. \end{aligned}$$

Magnetically conducting boundary:

$$\begin{aligned} H\text{-case} \Rightarrow\ & \varepsilon_1' P_{\nu-1/2}'(\cos\theta_0) P_{\nu-1/2}(-\cos\theta_0) \\ & + \varepsilon_2' P_{\nu-1/2}'(-\cos\theta_0) P_{\nu-1/2}(\cos\theta_0) = 0 \\ E\text{-case} \Rightarrow\ & \mu P_{\nu-1/2}(-\cos\theta_0) P_{\nu-1/2}'(\cos\theta_0) \\ & + \mu^* P_{\nu-1/2}(\cos\theta_0) P_{\nu-1/2}'(-\cos\theta_0) = 0 \end{aligned}$$

Perfectly conducting boundary:

$$\begin{aligned} H\text{-case} &\Rightarrow P_{\nu-1/2}(\cos\theta_0) = 0 \\ E\text{-case} &\Rightarrow P_{\nu-1/2}'(\cos\theta_0) = 0. \end{aligned}$$

It is obvious that the problem of finding the solutions to the transcendental equations obtained above are not exactly solvable. However a good approximate value for the minimal ν can be obtained by some appropriate numerical techniques which will be explained in the next section. Here we confine ourselves to notice only some obvious points.

i. in the case of (*impenetrable*) perfectly conducting boundaries considered above the minimal values of ν depend only on the apex angle θ_0 while

ii. in the cases of penetrable boundaries the values of ν depend also on the constitutive parameters ε_1', ε_2', μ and μ^* of the space in addition to the apex angle θ_0.

7.7 NUMERICAL SOLUTIONS OF THE TRANSCENDENTAL EQUATIONS SATISFIED BY THE MINIMAL INDEX

As mentioned above, the roots of the equations $\Delta_{ij}(\nu) = 0$ derived in the previous section cannot be found exactly. However, by dwelling on the idea used in Idemen [69] to determine the poles numerically, one can find a good approximate numerical value for the minimal ν in the case of general complex-valued coefficients $\varepsilon'_{1,2}$ and A. The main philosophy as well as the essential steps of this method will be explained in Section 7.7.3 below. However, if the tip is very sharp or the aforementioned coefficients are all real, then the minimal ν can be determined rather easily sometimes by using asymptotically valid analytic formulas or graphical methods. Now we will explain these methods.

7.7.1 The Case of Very Sharp Tip

When the tip is very sharp (i.e., when θ_0 is very close to π) and the resulting equation consists only of $P_{\nu-1/2}(\pm\cos\theta_0) = 0$ or of their product, we can use the approximate formulas to determine the minimal value of ν. Note that the formula (7.16a) is not suitable for this purpose because it gives in the first approximation $\nu = 1/2$, which contradicts (7.19). More correct values of ν—that is, the values in the second-order approximation—can be obtained from the fact that the equations $P_{\nu-1/2}(\pm\cos\theta_0) = 0$ have infinitely many zeros that are all simple, real, and symmetrically located with respect to the origin [72, p. 1030, no. 8.781] and have the following asymptotic expressions [72, p. 1030, no. 8.787]:

$$P_{\nu-1/2}(\cos\theta_0) = 0 \Rightarrow \nu \sim -\frac{1}{2} + k + 1/\left\{2\log\frac{2}{\pi-\theta_0}\right\}, \; k=1,2,\ldots \tag{7.38a}$$

$$P_{\nu-1/2}(-\cos\theta_0) = 0 \Rightarrow \nu \sim x_k/\{2\sin[(\pi-\theta_0)/2]\}, \tag{7.38b}$$

where $k = 1, 2, \ldots,$ and x_k refers to positive zeros of the zeroth order Bessel function, that is, $J_0(x_k) = 0$. The smallest positive values correspond to $k = 1$, namely,

$$P_{\nu-1/2}(\cos\theta_0) = 0 \Rightarrow \nu \sim \frac{1}{2} + \delta, \quad \delta = 1/\left\{2\log\frac{2}{\pi-\theta_0}\right\} > 0, \tag{7.39a}$$

$$P_{\nu-1/2}(-\cos\theta_0) = 0 \Rightarrow \nu \sim 1.2024/\sin[(\pi-\theta_0)/2], \tag{7.39b}$$

with the value of $x_1 = 2.4048$ being taken into account. It is obvious that the values determined through (7.39a) are always lesser than those given by (7.39b).

Consider, for example, the H-case polarization and *perfectly conducting* boundary. In this case from the equation obtained in Section 7.6 one writes

$$E_\phi \equiv 0, \quad E_{r,\theta} = O(r^{-1+\delta}), \quad H_{r,\theta} \equiv 0, \quad H_\phi = O(r^\delta).$$

Here δ refers to the small positive number given in (7.39a).

Consider, for example, the case when $\theta_0 = (11/12)\pi$ (= 165 degrees). In this case (7.39a) gives $\delta \approx 0,246$ and $\nu \approx 0,746$.

7.7.2 The Case of Real-Valued Minimal ν

Observe that the equations $\Delta_{ij}(\nu) = 0$ are all special cases of the following equation:

$$\begin{aligned} &a\sqrt{2}/\pi P_{\nu-1/2}(\cos\theta_0) \\ &\quad + b\sqrt{2}/\pi P'_{\nu-1/2}(\cos\theta_0) \\ &\quad + cP_{\nu-1/2}(\cos\theta_0)P_{\nu-1/2}(-\cos\theta_0) \\ &\quad + (d + e\sin\theta_0)P_{\nu-1/2}(\cos\theta_0)P'_{\nu-1/2}(-\cos\theta_0) \\ &\quad + (g + h\sin\theta_0)P'_{\nu-1/2}(\cos\theta_0)P_{\nu-1/2}(-\cos\theta_0) = 0. \end{aligned} \tag{7.40}$$

Here a, b, c, d, e, g, and h stand for appropriate constants. They all are real when $\varepsilon'_{1,2}$ and A, which appear in the equations $\Delta_{ij}(\nu) = 0$, are real. In such a case, one plots easily the graph of the function $\Delta_{ij}(\nu)$, whence one can determine the minimal root ν under the assumption that it is real. To simplify more the computations, in what follows we will use also the identity [72, p. 1021, no. 8.731(1)]

$$(z^2-1)P'_\nu(z) = (\nu+1)P_{\nu+1}(z) - (\nu+1)zP_\nu(z)$$

and define

$$f(\nu,\theta) = \int_0^\theta \frac{\cos\nu t}{\sqrt{\cos t - \cos\theta}}dt, \qquad 0 < \theta < \pi, \tag{7.41}$$

which yield [72, p. 1017, no. 8.714]

$$\begin{aligned}
P_{\nu-1/2}(\cos\theta_0) &= \frac{\sqrt{2}}{\pi} f(\nu,\theta_0),\\
P_{\nu-1/2}(-\cos\theta_0) &= \frac{\sqrt{2}}{\pi} f(\nu,\pi-\theta_0),\\
P'_{\nu-1/2}(\cos\theta_0) &= \frac{\sqrt{2}}{\pi}\frac{\nu+1/2}{\sin^2\theta_0}[\cos\theta_0 f(\nu,\theta_0) + f(\nu+1,\theta_0)],\\
P'_{\nu-1/2}(-\cos\theta_0) &= \frac{\sqrt{2}}{\pi}\frac{\nu+1/2}{\sin^2\theta_0}[\cos\theta_0 f(\nu,\pi-\theta_0)\\
&\qquad\qquad + f(\nu+1,\pi-\theta_0)]
\end{aligned}$$

and reduce the equation (7.40) into $H(\nu) = 0$ with

$$\begin{aligned}
H(\nu) = &\left[a + b(\nu+1/2)\frac{\cos\theta_0}{\sin^2\theta_0}\right] f(\nu,\theta_0)\\
&- b(\nu+1/2)\frac{1}{\sin^2\theta_0} f(\nu+1,\theta_0)\\
&+ [c + b(\nu+1/2)\frac{\cos\theta_0}{\sin^2\theta_0}\{(g-d)\\
&\qquad + (h-e)\sin\theta_0\}] f(\nu,\theta_0) f(\nu,\pi-\theta_0)\\
&- (\nu+1/2)\frac{d + e\sin\theta_0}{\sin^2\theta_0} f(\nu,\theta_0) f(\nu+1,\pi-\theta_0)\\
&- (\nu+1/2)\frac{g + h\sin\theta_0}{\sin^2\theta_0} f(\nu+1,\theta_0) f(\nu,\pi-\theta_0).
\end{aligned} \tag{7.42}$$

Thus from the graph of the function $H(\nu)$ one can find the value of the minimal ν.

As simple instructive examples we consider here the particular cases where one has

$$\begin{aligned}
&\text{Case-A}: && \theta_0 = (11/12)\pi, \quad c = 1 \\
& && \text{while } a = b = d = e = g = h = 0, \\
&\text{Case-B}: && \theta_0 = (11/12)\pi, \quad d = g = 1 \\
& && \text{while } a = b = c = e = h = 0, \\
&\text{Case-C}: && \theta_0 = (11/12)\pi, \quad d = 0,35, \quad g = 1 \\
& && \text{while } a = b = c = e = h = 0.
\end{aligned}$$

Note that the first case concerns the H case with resistive and transmissive ($p \neq 0$) boundaries as well as the E case with magnetically conducting and transmissive ($n \neq 0$) boundaries while the other two cases concern the simple interface (E and H cases) as well as the resistive (E case), modified conducting (E case) and magnetically conducting (H case) boundaries. In the second case $\varepsilon_2'/\varepsilon_1' = 1$ or $\mu_2/\mu_1 = 1$ depending on the H or E case, while in the third case $\varepsilon_2'/\varepsilon_1' = 0.35$ or $\mu_2/\mu_1 = 0.35$ depending on the H or E case.

Figure 7.2 shows the graph pertinent to case A. It is interesting to observe that the minimal root, which is nearly equal to $v = 0,76$, is very close to the value obtained in Section 7.7.1 above by the formula (7.39a). The graphs corresponding to cases B and C are shown in Fig. 7.3 and 7.4. From these figures one observes that $v \approx 0.770$ is a simple root for the B case while $v \approx 0.698$ is a double root for the C case. Therefore in the latter case the singularity is logarithmic.

7.7.3 A Function-Theoretic Method to Determine Numerically the Minimal *v*

As already mentioned, the numerical methods given in Section 7.7.2 are not effective when the coefficients taking place in Δ_{ij} are complex. Now, by dwelling on the idea used in Idemen and Akduman [74] (see also Idemen [69]), we will try to establish a method that enables us to determine the minimal value in question if it is *simple* and $\Delta_{ij}(v)$ has no other zero with the same real part. For the sake of simplicity, in what follows we will omit the sub-indices (ij) existing in $\Delta_{ij}(v)$.

Our starting point is the fact that Legendre functions and their derivatives appearing in $\Delta(v)$ are all entire functions of v. This can

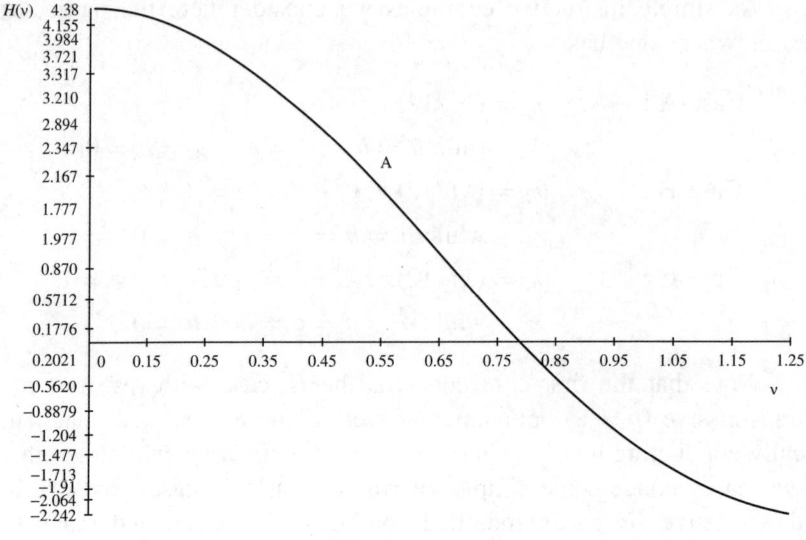

Figure 7.2. Graph of the function $H(\nu)$ for the case A.

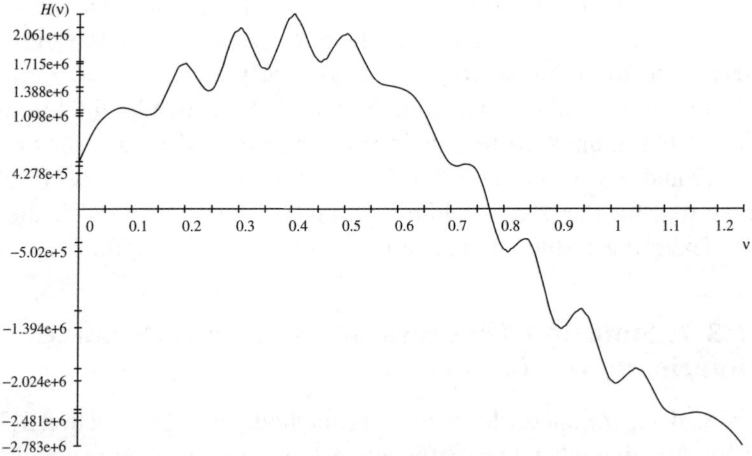

Figure 7.3. Graphs of the function $H(\nu)$ for case B.

easily be seen, for example, from the expression (see Gradshteyn and Ryzhik [72, p.1017, no. 8.714])

$$P_{\nu-1/2}(\cos\theta) = \frac{\sqrt{2}}{\pi}\int_0^\theta \frac{\cos \nu t}{\sqrt{\cos t - \cos\theta}}dt, \qquad 0 < \theta < \pi. \quad (7.43)$$

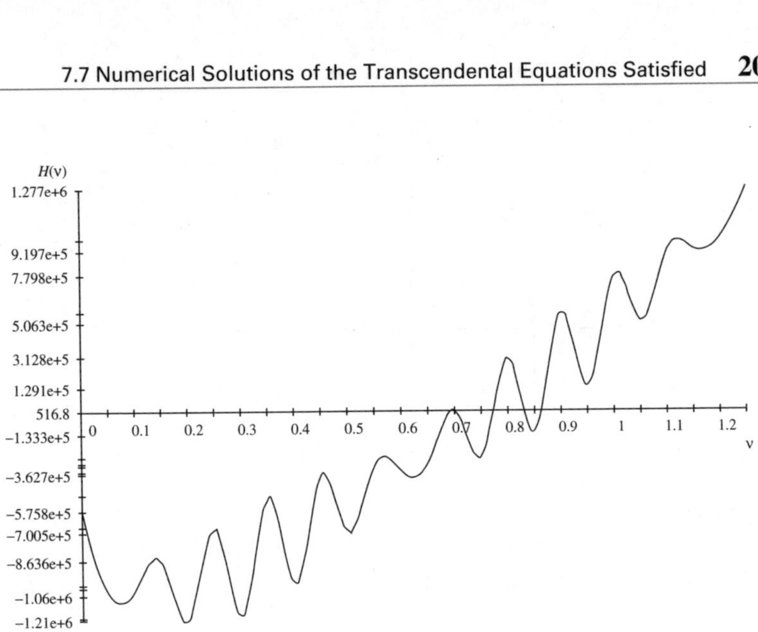

Figure 7.4. Graphs of the function $H(v)$ for case C.

Indeed, this integral converges uniformly for all finite v, which ensures the entirety of $P_{v-/2}(\cos\theta)$ [38, p. 217]. Therefore, the function $f(v) = \Delta'(v)/\Delta(v)$ consists of a meromorphic function that has simple poles at the zeros of $\Delta(v)$, say $v_m(m = 1, 2, \ldots)$. If the multiplicity of v_m is equal to N_m, then the residue of the function $f(v)$ at the pole $v = v_m$ is equal to N_m.

Now assume that $\Delta(v)$ has no zero on the imaginary axis and define

$$A_\alpha = \frac{1}{2\pi i} \fint_{C_\alpha} \frac{\Delta'(v)}{\Delta(v)} \frac{dv}{v}, \tag{7.44}$$

where C_α stands for the vertical straight line passing through the point $v = \alpha \geq 0$ while the horizontal bar on the integral sign refers to the Cauchy principal value (see Fig. 7.5). Then one has

$$A_\alpha - A_0 = \sum_{\Re v_m \in (0,\alpha)} \frac{N_m}{v_m} + \frac{1}{2} \sum_{\Re v_m = \alpha} \frac{N_m}{v_m} + \frac{1}{2} \frac{\Delta'(0)}{\Delta(0)}, \quad \alpha > 0. \tag{7.45}$$

If between the lines C_0 and C_α there is only one zero of $\Delta(v)$, namely the minimal v, then (7.45) becomes

$$A_\alpha - A_0 = \frac{1}{v} + \frac{1}{2} \frac{\Delta'(0)}{\Delta(0)} \tag{7.46}$$

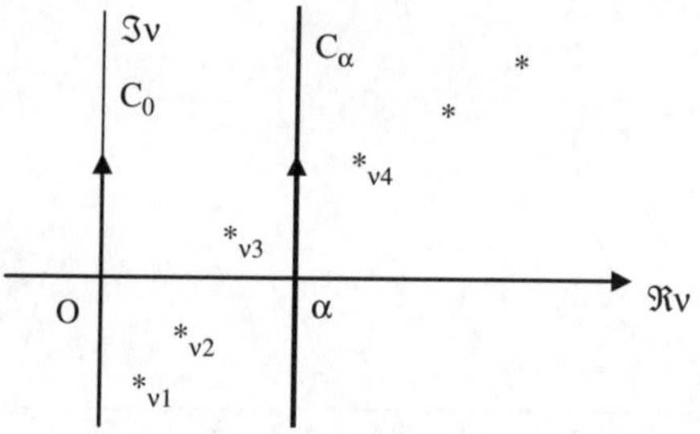

Figure 7.5. The zeros of the functions $\Delta_{ij}(\nu)$

which yields

$$\nu = 2 \Big/ \left[2A_\alpha - 2A_0 - \frac{\Delta'(0)}{\Delta(0)} \right]. \tag{7.47}$$

Equation (7.47) hints the following steps:

1. Compute first A_0 and $\Delta'(0)/\Delta(0)$.
2. Then choose a small $\alpha > 0$ and compute A_α.
3. Compute the real part of $[2A_\alpha - 2A_0 - \Delta'(0)/\Delta(0)]$.

If the latter is equal to zero, then $\Re\nu > \alpha$. In this case one repeats the steps 2 and 3 with a larger α. Conversely, if the aforementioned real part is positive, then one has $\Re\nu \leq \alpha$. In this case one repeats the steps 2 and 3 with a smaller α. By this kind of successive iteration, one can arrive at such a position that inside the strip $0 < \Re\nu < \alpha$ there is only one zero, namely ν given by (7.47). After having found a ν given by (7.47), it is an easy matter to check if it is exactly a simple zero of $\Delta(\nu)$.

PROBLEMS

1. Let S be a modified conducting planar sheet on which the boundary conditions of the following types are satisfied (see Section 6.5):

$$\mathbf{E}_{1t} - \mathbf{E}_{2t} + A \operatorname{grad}(E_{1n} + E_{2n}) = 0,$$

$$\mathbf{e}_n \times (\mathbf{H}_{1t} - \mathbf{H}_{2t}) + \frac{1}{2R}(\mathbf{E}_{1t} + \mathbf{E}_{2t}) = 0.$$

Here R stands for a given positive constant while A is a given complex-valued parameter.

(a) Show that the boundary condition table is as follows:

$$C = \begin{bmatrix} 1 & -1 & 1/(2R) & 1/(2R) & 0 & 0 & 0 & 0 \\ 0 & 0 & 1 & -1 & 0 & 0 & A & A \end{bmatrix}.$$

(b) If the above-mentioned sheet becomes curved *conically* what will be the boundary condition table at points far away from the tip?.

(c) Try to find the boundary condition table near the tip and use the result to reveal the equation satisfied by the minimal exponent ν.

2. Consider pr.-1 for the case of a *general transmissive* boundary on which one has

$$\mathbf{E}_{1t} = m\mathbf{E}_{2t} + n\ \mathbf{e}_\theta \times \mathbf{H}_{2t}, \quad \mathbf{e}_\theta \times \mathbf{H}_{1t} = p\mathbf{E}_{2t} + q\ \mathbf{e}_\theta \times \mathbf{H}_{2t},$$

which yields

$$C = \begin{bmatrix} 0 & -n & 1 & -m & 0 & 0 & 0 & 0 \\ 1 & -q & 0 & -p & 0 & 0 & 0 & 0 \end{bmatrix}.$$

Here m, n, p and q are given complex-valued constants.

CHAPTER 8

Temporal Discontinuities

8.1 UNIVERSAL INITIAL CONDITIONS

Suppose that in the time interval $0 < t < \tau$ certain changes occur in some components of an electromagnetic field. In order to reveal the effect of these changes on the evolution of the field for $t > \tau$, one has to solve the Maxwell equations (completed with constitutive equations) under the *boundary* and *initial conditions*.* These latter require the knowledge of the value of the field for $t = \tau + 0$, which is not known beforehand inasmuch as we know only the field at $t = -0$. Hence the finding of the relations connecting the values at $t = \tau + 0$ and $t = -0$ constitutes a basic problem of vital importance. In many practical applications the duration τ is so short as compared to specific time intervals (for example, *period, relaxation time*, etc.) related to the field that the simplifications made by assuming $\tau \to 0$ give a good approximation. The shock effects in biological systems and thunderbolts constitute some simple examples of this situation. In such cases the aforementioned changes are described via Dirac's generalized function $\delta(t)$, which reduces the above-mentioned problem to that of finding the relations interrelating the field values at $t = -0$ and $t = +0$. Our aim is now to this end through the postulate (assumption) cited in Section 3.1.†

Assume that the charge densities as well as the field components of an electromagnetic field suffer some instantaneous changes at the

*To these conditions one has, sometimes, to add edge and radiation conditions also.
†See Idemen [19, Chapter 2].

Discontinuities in the Electromagnetic Field, First Edition. By M. Mithat Idemen.

Published 2011 by John Wiley & Sons, Inc.

time $t = 0$. Then, in accordance with Theorem 2.4 of Section 2.7, one has

$$\rho(\mathbf{r},t) = \{\rho(\mathbf{r},t)\} + \sum_{k=0} \rho_k(\mathbf{r})\delta^{(k)}(t), \tag{8.1a}$$

$$\mathbf{J}(\mathbf{r},t) = \{\mathbf{J}(\mathbf{r},t)\} + \sum_{k=0} \mathbf{J}_k(\mathbf{r})\delta^{(k)}(t), \tag{8.1b}$$

$$\mathbf{E}(\mathbf{r},t) = \{\mathbf{E}(\mathbf{r},t)\} + \sum_{k=0} \mathbf{E}_k(\mathbf{r})\delta^{(k)}(t). \tag{8.1c}$$

(similar expressions for other quantities)

If we insert these expressions into the Maxwell equations reduced to the vacuum (see (3.1.a–d) and (3.3)) and use the relation

$$\frac{\partial}{\partial t}\{f(t)\} = \left\{\frac{\partial}{\partial t}f(t)\right\} + [[f]]\delta(t), \tag{8.2}$$

then the singular parts of order zero result in the following equations:

$$\text{curl}\mathbf{H}_0 - \varepsilon_0[[\mathbf{E}]] = \mathbf{J}_0 + [[\mathbf{P}]], \tag{8.3a}$$

$$\text{curl}\mathbf{E}_0 + \mu_0[[\mathbf{H}]] = -[[\mathbf{M}]], \tag{8.3b}$$

$$\varepsilon_0\text{div}\mathbf{E}_0 = \rho_0 - \text{div}\mathbf{P}_0, \tag{8.3c}$$

$$\mu_0\text{div}\mathbf{H}_0 = -\text{div}\mathbf{M}_0, \tag{8.3d}$$

$$\text{div}\mathbf{J}_0 + [[\rho]] = 0, \tag{8.3e}$$

while the higher-order singular parts give (in what follows $k = 1,2,\ldots$)

$$\text{curl}\mathbf{H}_k - \varepsilon_0\mathbf{E}_{k-1} = \mathbf{J}_k + \mathbf{P}_{k-1}, \tag{8.4a}$$

$$\text{curl}\mathbf{E}_k + \mu_0\mathbf{H}_{k-1} = -\mathbf{M}_{k-1}, \tag{8.4b}$$

$$\varepsilon_0\text{div}\mathbf{E}_k = \rho_k - \text{div}\mathbf{P}_k, \tag{8.4c}$$

$$\mu_0\text{div}\mathbf{H}_k = -\text{div}\mathbf{M}_k, \tag{8.4d}$$

$$\text{div}\mathbf{J}_k + \rho_{k-1} = 0. \tag{8.4e}$$

The relations (8.3a,b) and (8.3e), which involve the jumps $[[\cdot]]$, give the *temporal discontinuities* (i.e., the initial conditions) in their most general forms. Hence we can call them *universal initial conditions*. By introducing the usual definitions $\mathbf{P} + \varepsilon_0\mathbf{E} = \mathbf{D}$ and $\mathbf{M} +$

$\mu_0 \mathbf{H} = \mathbf{B}$, they can also be written as follows:

$$[[\mathbf{D}] = -\mathbf{J}_0 + \text{curl}\mathbf{H}_0, \tag{8.5a}$$
$$[[\mathbf{B}]] = -\text{curl}\mathbf{E}_0, \tag{8.5b}$$
$$[[\rho]] = -\text{div}\mathbf{J}_0. \tag{8.5c}$$

The remaining relations (8.3c,d) and (8.4a–e) are merely the *compatibility conditions* to be satisfied by singular terms of different orders. They can be rearranged as follows (in what follows $k = 0, 1, \ldots$):

$$\text{curl}\mathbf{H}_{k+1} - \mathbf{D}_k = \mathbf{J}_{k+1}, \tag{8.6a}$$
$$\text{curl}\mathbf{E}_{k+1} + \mathbf{B}_k = 0, \tag{8.6b}$$
$$\text{div}\mathbf{D}_k - \rho_k = 0, \tag{8.6c}$$
$$\text{div}\mathbf{B}_k = 0, \tag{8.6d}$$
$$\text{div}\mathbf{J}_{k+1} + \rho_k = 0. \tag{8.6e}$$

8.2 LINEAR MEDIUMS IN THE GENERALIZED SENSE

The current density $\mathbf{J} = \{\mathbf{J}\} + \sum \mathbf{J}_k \delta^{(k)}(t)$ appearing above is composed, in general, of two terms. One of these terms, say $\mathbf{J}_c$, is created by the conduction of the medium, if any, while the other, say $\mathbf{J}_v$, is related to the motion of free charges. If the conduction current density $\mathbf{J}_c$ as well as the fields $\mathbf{D}$ and $\mathbf{B}$ can be written as

$$\mathbf{D} = \varepsilon \mathbf{E}, \qquad \mathbf{B} = \mu \mathbf{H}, \qquad \mathbf{J} = \sigma \mathbf{E}, \tag{8.7}$$

where ε, μ, and σ stand for given scalar or tensor-valued functions of $\mathbf{r}$ and t, then the medium in which the field propagates is said to be *linear in the generalized sense*. For such a medium the following theorem is valid.

THEOREM 8.1

If there exists an integer $k_0 \geq 0$ such that $\mathbf{J}_{vk} = 0$ for all $k \geq k_0{+}1$, then in a linear (in the generalized sense!) medium one has

$$\mathbf{E}_k = \mathbf{D}_k = \mathbf{H}_k = \mathbf{B}_k = \rho_k = 0, \qquad k = k_0, k_0 + 1, \ldots. \tag{8.8}$$

This theorem shows how we can elaborate the compatibility conditions. It can be proved by considering the fact that the number of the singular terms appearing in (8.1a,b, ...) (i.e., the order of the distributions pertinent to the field in question) cannot be infinite. Hence, there is a finite number $k_1 > k_0$ such that all the singular terms with sub-index k_1 are zero. Then, by considering the conditions cited in the theorem and the relations (8.6a–e), we conclude that (8.8), which is satisfied for $k = k_1$, is also satisfied for $k = k_1 - 1$. This reduction process can obviously be repeated until $k = k_0$, which proves the theorem. Note that the process cannot be forwarded beyond $k = k_0$ because J_{vk_0} is different from zero.

Suppose, for example, that $k_0 = 0$ and

$$\mathbf{J}_v(\mathbf{r}, t) = \{\mathbf{J}_v(\mathbf{r}, t)\} + \mathbf{J}_{v0}(\mathbf{r})\delta(t). \tag{8.9a}$$

Then, from (8.8) and (8.5a–d) we get

$$\mathbf{E}_k = \mathbf{D}_k = \mathbf{H}_k = \mathbf{B}_k = \rho_k = 0, \qquad k = 0, 1, \ldots \tag{8.9b}$$

and

$$[[\mathbf{D}]] = -\mathbf{J}_{v0}(r), \qquad [[\mathbf{B}]] = 0, \qquad [[\rho]] = -\text{div}\mathbf{J}_{v0}(r). \tag{8.9c}$$

8.3 AN ILLUSTRATIVE EXAMPLE

Suppose that in an infinite medium filled with a simple material an instantaneous discharge takes place at $t = 0$ in the plane $x = 0$, which causes a displacement of charge of amount Q coulomb/m in a certain direction **e**. The problem is the determination of the wave created by this discharge.

To solve this problem, consider first the fact that the density of the current related to the above-mentioned discharge, namely

$$\mathbf{J}_v = Q\delta(x)\delta(t)\mathbf{e} \qquad (\text{Amp/m}^2),$$

is a function of x and t only, which enables us to write

$$\mathbf{E} = [g_1(x-ct) + h_1(x+ct)]\mathbf{e}_y + [g_2(x-ct) + h_2(x+ct)]\mathbf{e}_z, \quad t > 0$$

$$\sqrt{\tfrac{\mu}{\varepsilon}}\mathbf{H} = [g_1(x-ct) - h_1(x+ct)]\mathbf{e}_z - [g_2(x-ct) - h_2(x+ct)]\mathbf{e}_y, \quad t > 0.$$

Now let us put these expressions into (8.9c) and consider the fact that $\mathbf{E} \equiv \mathbf{0}, \mathbf{H} \equiv \mathbf{0}$ for $t = -0$. Thus we can easily find

$$g_1(x) = h_1(x) = -\frac{Q}{2\varepsilon}(\mathbf{e} \cdot \mathbf{e}_y)\delta(x),$$

$$g_1(x) = h_1(x) = -\frac{Q}{2\varepsilon}(\mathbf{e} \cdot \mathbf{e}_z)\delta(x),$$

which finally give

$$\mathbf{E} = -\frac{Q}{2\varepsilon}[\delta(x - ct) + \delta(x + ct)]\mathbf{e}, \qquad t > 0$$

and

$$\mathbf{H} = -\frac{Qc}{2}[\delta(x - ct) - \delta(x + ct)]\mathbf{e_x} \times \mathbf{e}, \qquad t > 0.$$

PROBLEM

1. The basic equation of the photoacoustic and thermoacoustic tomographies is as follows:

$$\Delta p(\mathbf{x}, t) - \frac{1}{c^2}\frac{\partial^2}{\partial t^2}p(\mathbf{x}, t) = -p_0(\mathbf{x})\delta'(t), \quad \mathbf{x} \in D \subset R^3, \quad t \in (-\infty, \infty).$$

Here $p(\mathbf{x}, t)$ denotes the pressure field excited by an electromagnetic pulse illumination in a soft tissue that fills region D. Show that if $p(\mathbf{x}, t) \equiv 0$ before the illumination, then it satisfies the following initial-value problem:

$$\Delta p(\mathbf{x}, t) - \frac{1}{c^2}\frac{\partial^2}{\partial t^2}p(\mathbf{x}, t) = 0, \qquad \mathbf{x} \in D, \quad t > 0$$

$$p(\mathbf{x}, 0) = c^2 p_0(\mathbf{x}), \quad \frac{\partial p}{\partial t}(\mathbf{x}, 0) = 0, \qquad \mathbf{x} \in \mathrm{D}.$$

References

1. C. J. Bouwkamp, A note on singularities occuring at sharp edges in electromagnetic diffraction theory, *Physica*, Vol. 12, pp. 467–475, 1946.
2. W. I. Smirnow, *Lehrgang der höheren Mathematik*, Teil II. VEB Deutscher Verlag, Berlin, 1968.
3. D. S. Jones, *The Theory of Electromagnetism*, Pergamon Press, London, 1964, p. 567.
4. S. A. Schelkunoff, On teaching the undergraduate electromagnetic theory, *IEEE Trans. Ed.*, Vol. E-15, Section 5, pp. 15–25, 1972.
5. M. Idemen, The Maxwell equations in the sense of distributions, *IEEE Trans. Antennas Propagat.*, Vol. 21, pp. 736–738, 1973.
6. G. R. Kirchhoff, *Sitz. Preuss. Akad. Wiss., Berlin. See also: Vorlesungen über Mathematische Physik* (2), 22, Teubner, Leipzig, 1877.
7. P. A. M. Dirac, *The Principles of Quantum Mechanics*, 4th edition (revised), Clarendon Press, Oxford, 1967, p. 58.
8. L. Schwartz, *Théorie des Distributions*, tome 1 (1950), tome 2 (1951) I Hermann, Paris.
9. R. A. Adams, *Sobolev Spaces*, Academic Press, New York, 1975.
10. I. M. Gel'fand and G. E. Shilov, *Generalized Functions*, Vol. 2, Academic Press, New York, 1968.
11. K. Yoshida, *Functional Analysis*, Springer-Verlag, Berlin, 1966.
12. E. T. Whittaker and G. N. Watson, *A Course of Modern Analysis Cambridge*, University Press, Cambridge, 1958.
13. J. C. Maxwell, *Treatise on Electricity and Magnetism*, 1873.
14. A. M. Ampère, *Annales de chimie et de physique* (15), 1820.
15. M. Faraday, *Diary*, Royal Institution, London, 1932. See also: G. Gamow, *Biography of Physics*, Harper and Row, New York, 1961.
16. M. Idemen, Universal boundary relations of the electromagnetic field, *J. Phys. Soc. Japan*, Vol. 59, pp. 71–80, 1990.
17. A. R. Panicali, On the boundary conditions at surface current distributions described by the first derivative of Dirac's impulse, *IEEE Trans. Antennas Propagat.*, Vol. 25, pp. 901–903, 1977.

Discontinuties in the Electromagnetic Field, First Edition. By M. Mithat Idemen.

Published 2011 by John Wiley & Sons, Inc.

18. J. Van Bladel, Boundary conditions at double current sheet, *Electron. Lett.*, Vol. 25, pp. 98–99, 1989.
19. M. Idemen, Universal boundary conditions and Cauchy data for the electromagnetic field, in *Essays on the Formal Aspects of Electromagnetic Theory*, pp. 657–698, A. Lakhtakia (ed.), World Sci., N.J., 1993.
20. H. Bateman, *Electrical and Optical Wave Motion*, Cambridge University Press, London, 1915.
21. R. F. Harrington and J. R. Mautz, An impedance sheet approximation for thin dielectric shells, *IEEE Trans. Antennas Propagat.*, vol. 23, pp. 531–534, 1975.
22. T. B. A. Senior, Scattering by resistive strips, *Radio Sci.*, vol. 14, p. 911–924, 1979.
23. T. B. A. Senior, Combined resistive and conducting sheets, *IEEE Trans. Antennas and Propagat.*, vol. AP-33, p. 577–579, 1985.
24. T. B. A. Senior and J. L. Volakis, Sheet simulation of a thin dielectric layer, *Radio Sci.*, vol. 22, p. 1261–1272, 1987.
25. E. L. Feinberg, On the propagation of radio waves along an imperfect surface, *J. Phys. (USSR)*, p. 317–330, 1944.
26. V. A. Fock, Solution of the problem of propagation of electromagnetic waves along the Earth's surface by method of parabolic equation, *J. Phys. (USSR)*, pp. 13–24, 1946.
27. M. A. Leontovich, *Investigations on Radiowave Propagation*, Part 2, Academy of Sciences, Moscow, 1948.
28. M. Idemen, Necessary and sufficient conditions for a surface to be an impedance boundary, *AEÜ (Electron. Commun.)*, Vol. 35, pp. 84–86, 1981.
29. S. W. Lee and W. Gee, How good is the impedance boundary condition, *IEEE Trans. Antennas Propagat.*, vol. 35, pp. 1313–1315, 1987.
30. J. R. Wait, Exact surface impedance for a cylindrical conductor, *Electron. Lett.*, vol. 15, pp. 659–660, 1979.
31. J. R. Wait, Exact surface impedance for a spherical conductor, *Proc. IEEE*, vol. 68, pp. 279–281, 1980.
32. H. A. Lorentz, Electromagnetic phenomena in a system moving with any velocity smaller than that of light, *Proc. R. Acad. Amsterdam*, pp. 809–831, 1904.
33. H. Poincaré, Sur la dynamique de l'electron, *Rend Circ. Mat. Palermo*, vol. 21, pp. 129–175, 1906.
34. A. Einstein, Zur Elektrodynamik bewegter Körper, *Ann. Phys.*, 891–921, 1905.
35. W. Ignatowski, Das Relativitätsprinzip, *Arch. Math. Phys.*, vol. 17, pp. 1–24, 1910.
36. A. A. Robb, *A Theory of Space and Time*, Cambridge University Press, Cambridge, 1914.
37. M. Idemen, Derivation of the Lorentz transformation from the Maxwell equations, *J. Electromagn. Waves Appl. (JEMWA)*, vol. 19(4), pp. 451–467, 2005.

38. W. I. Smirnow, *Lehrgang der höheren Mathematik*, Teil III. 2, 2. Auflage, p: 55, VEB Deutscher Verlag der Wissenschaften, Berlin, 1959.
39. V. Namias, *Am. J. Phys*. vol. 56, pp. 898–904, 1988.
40. A. Sommerfeld, *Elektrodynamik*, Wiesbaden, Germany i Diterichsche Verlag, 1948, pp. 281–289.
41. H. Epheser and T. Schlomka, Flachengrössen und elektrodynamische Grentzbedingungen bei bewegter Körpern, *Ann. Phys*., vol. 6, pp. 211–220, 1951.
42. M. Van Laue, *Die Relativitatstheorie*, erster Band, Die spezielle Relativitatstheorie, Viewig, 1955.
43. H. Arzeliès and J. Henry, *Milieux Conducteurs ou Polarisables en Mouvement*, Gauthier-Villars, Paris, 1951, pp. 34–35, 45, 132–163, 171–178, 291.
44. R. C. Costen and D. Adamson, Three dimensional derivation of the electrodynamic jump conditions and momentum-energy laws at a moving boundary, *Proc. IEEE*, vol. 53, no. 9, pp. 1181–1196, 1965.
45. J. Meixner, Die Kantenbedingung in der Theorie der Beugung Elektromagnetischer Wellen an vollkommen leitenden ebenen Schirmen, *Ann. Phys*., 6, pp. 2–9, 1949.
46. W. Braunbek, On the diffraction field near a plane-screen corner, *IRE Trans. Antennas Propagat*., vol. AP-4, pp. 219–223, 1956.
47. P. Poincelot, Généralisation de la condition aux arêtes dans le cas d'une courbe gauche, *C. R. Acad. Sci. Paris*, vol. 251, pp. 2670–2671, 1960.
48. R. A. Hurd, On the possibility of intrinsic loss occuring at the edges of ferrites, *Can. J. Phys*., vol. 40, pp. 1067–1076, 1962.
49. R. A. Hurd, Intrinsic loss at the edges of anisotropic plasmas, *Can. J. Phys*., vol. 41, pp. 1554–1562, 1963.
50. R. Mittra and S. W. Lee, Edge condition and intrinsic loss in uniaxial plasma, *Can. J. Phys*., vol. 46, pp. 111–120, 1968.
51. R. Mittra and S. W. Lee, *Analytical Techniques in the Theory of Guided Waves*, Macmillan, New York, 1971.
52. J. Meixner, The behavior of electromagnetic fields at edges, *IEEE Trans. Antennas Propagat*., vol. AP-20, pp. 442–446, 1972.
53. K. C. Lang, Edge condition of a perfectly conducting wedge with its exterior region divided by a resistive sheet, *IEEE Trans. Antennas Propagat*., pp. 237–238, 1973.
54. R. A. Hurd, The edge condition in electromagnetics, *IEEE Trans. Antennas Propagat*., pp. 70–73, 1976.
55. J. B. Andersen and V. V. Solodukhov, Field behavior near a dielectric wedge, *IEEE Trans. Antennas Propagat*., vol. AP-26, pp. 598–602, 1978.
56. R. A. Hurd, On Meixner's edge condition for dielectric edges, *Can. J. Phys*., vol. 55, pp. 1970–1971, 1977.

57. I. M. Braver, P. S. Fridberg, K. L. Garb, and I. M. Yakover, The behavior of the electromagnetic field near the edge of resistive half-plane, *IEEE Trans. Antennas Propagat.*, vol. AP-36, pp. 1760–1768, 1988.
58. J. V. Bladel, Field singularities at metal–dielectric wedge, *IEEE Trans Antennas Propagat.*, vol. AP-33, pp. 450–455, 1985.
59. M. Idemen, Confluent edge conditions for the electromagnetic wave at the edge of a wedge bounded by material sheets, *Wave Motion*, vol. 32, no. 1, pp. 37–55, 2000.
60. A. W. Maue, Zur Formulierung eines allgemeinen Beugungsproblems durch eine Integralgleichung, *Z. Phys.*, vol. 126, p. 601, 1949.
61. D. S. Jones, Note on diffraction by an edge, *Q. J. Mech.*, vol. 3, pp. 420–434, 1950.
62. E. C. Titchmarsh, *Eigenfunction Expansions*, Oxford Clarendon Press, Oxford, 1962.
63. R. F. Harrington, Time-harmonic electromagnetic Field, McGraw-Hill, New York, 1961, pp. 199–202.
64. F. G. Leppington, Travelling waves in s dielectric slab with an abrupt change in thickness, *Proc. R. Soc. (London)*, 386/A, pp. 443–460, 1983.
65. J. van Bladel, Field singularity at the tip of a dielectric cone, *IEEE Trans. Antennas Propagat.*, vol. AP-33, pp. 893–894, 1985.
66. F. Olyslager, Overview of the singular behaviour of electromagnetic fields at edge and tips in bi-isotropic and special bi-anisotropic media, *Radio Sci.*, vol. 30, pp. 1349–1354, 1995.
67. D. S. Jones, Scattering by a cone, *Q. J. Mech. Appl. Math.*, vol. 50, pp. 499–523, 1997.
68. J. M. L. Bernard and M. A. Lyalinov, Diffraction of scalar waves by an impedance cone of arbitrary cross section, *Wave Motion*, vol. 33, pp. 155–181, 2001.
69. M. Idemen, Confluent tip singularity of the electromagnetic field at the apex of a material cone, *Wave Motion*, vol. 38(3) pp. 251–277, 2003.
70. I. H. Sneddon, *The Use of Integral Transforms*, McGraw-Hill, New York, 1972.
71. D. S. Jones, The Kontotovich–Lebedev transform, *J. Inst. Math. Applics.*, vol. 26, pp. 133–141, 1980.
72. I. S. Gradshteyn and I. M. Ryzhik, *Table of Integrals, Series, and Products*, Academic Press, New York, 1994.
73. G. K. Carrier, M. Krook, and C. E. Pearson, *Functions of a Complex Variable*, McGraw-Hill, New York, 1966.
74. M. Idemen and I. Akduman, Some geometrical inverse problems connected with two-dimensional static fields, *SIAM J. Appl. Math.*, vol. 48(3), pp. 703–718, 1988.
75. M. Idemen, Special theory of relativity and universal boundary conditions on uniformly moving surfaces, in *Proceedings 12th International Conference on MMET* (*Mathematical Methods in Electromagnetic Theory*), June 29–July 2, 2008, Odessa (Ukraine), pp. 24–30.

Index

Discontinuties in the Electromagnetic Field, First Edition. By M. Mithat Idemen.

Published 2011 by John Wiley & Sons, Inc.

H

I

J

K

L

M

IEEE PRESS SERIES ON ELECTROMAGNETIC WAVE THEORY

Andreas C. Cangellaris, *Series Editor*
University of Illinois, Urbana-Champaign, Illinois

1. *Field Theory of Guided Waves, Second Edition*
   Robert E. Collin
2. *Field Computation by Moment Methods*
   Roger F. Harrington
3. *Radiation and Scattering of Waves*
   Leopold B. Felsen, Nathan Marcuvitz
4. *Methods in Electromagnetic Wave Propagation, Second Edition*
   D. S. J. Jones
5. *Mathematical Foundations for Electromagnetic Theory*
   Donald G. Dudley
6. *The Transmission-Line Modeling Method: TLM*
   Christos Christopoulos
7. *The Plane Wave Spectrum Representation of Electromagnetic Fields, Revised Edition*
   P. C. Clemmow
8. *General Vector and Dyadic Analysis: Applied Mathematics in Field Theory*
   Chen-To Tai
9. *Computational Methods for Electromagnetics*
   Andrew F. Peterson
10. *Plane-Wave Theory of Time-Domain Fields: Near-Field Scanning Applications*
    Thorkild B. Hansen
11. *Foundations for Microwave Engineering, Second Edition*
    Robert Collin
12. *Time-Harmonic Electromagnetic Fields*
    Robert F. Harrington
13. *Antenna Theory & Design, Revised Edition*
    Robert S. Elliott
14. *Differential Forms in Electromagnetics*
    Ismo V. Lindell
15. *Conformal Array Antenna Theory and Design*
    Lars Josefsson, Patrik Persson
16. *Multigrid Finite Element Methods for Electromagnetic Field Modeling*
    Yu Zhu, Andreas C. Cangellaris
17. *Electromagnetic Theory*
    Julius Adams Stratton
18. *Electromagnetic Fields, Second Edition*
    Jean G. Van Bladel

19. *Electromagnetic Fields in Cavities: Deterministic and Statistical Theories*
David A. Hill
20. *Discontinuities in the Electromagnetic Field*
M. Mithat Idemen
21. *Understanding Geometric Algebra for Electromagnetic Theory*
John W. Arthur
22. *The Power and Beauty of Electromagnetic Theory*
Frederic R. Morgenthaler

Forthcoming

Modern Lens Antennas for Communications Engineering
John Thornton, Kao-Cheng Huang